Ecce Fides

Pillar of Truth

Ideal for
RCIA
Adult and Youth Bible Study
Homeschooling
Catholic Identity Studies

Answers to
Protestantism & Judaism
Pseudo-Protestantism
Mormonism
Jehovah's Witnesses
Atheism & Agnosticism
Secularism
Occultism

Fr. John J. Pasquini

D1402277

This book is printed in the United States by the Shepherds of Christ.

ISBN Number: 978-1-934222-23-2

Printed with Ecclesiastical Approval, 2007

Cover Graphic Design: Catherine Lipps

Ecce Fides is the third revision of a previous work. The original work was an elementary outline of Catholic apologetics published as *Why Catholic?* by St. Bede's Press. Copyright was returned to the author, Fr. John J. Pasquini. A revised and improved version was later published as *Catholic Answers: Protestant Questions*, copyright Fr. John J. Pasquini. The following text is an improved and revised version of the original outlines of the previous works, copyright Fr. John J. Pasquini.

Bible quotations are primarily from the New American Bible, unless otherwise indicated. Translations from the Fathers of the Church are from the various editions of Harvard's Classical Loeb Series, unless otherwise noted.

Shepherds of Christ Publications
P.O. Box 627
China, Indiana 47250 USA

Tel: (812) 273-8405
Toll free: (888) 211-3041
Fax: (812) 273-3182
Email: info@sofc.org
http://www.sofc.org

DEDICATION

To my brother priests, Shepherds of Christ.

Every book purchased goes to support free spiritual newsletters, books, booklets and many other spiritual and intellectual aids for priests worldwide.

**Love your priests! Support your priests!
Nourish your priests!**

TABLE OF CONTENTS

FOREWARD

Anywhere from twenty to eighty percent of Protestant and pseudo-Christian denominations are made up of former Catholics. And sadly, there are too many practicing Catholics who do not believe in the Church's infallibility in the areas of faith and morals.

This book is intended to combat this dangerous trend. *Ecce Fides* is a book intended to reaffirm Catholics in their faith and in the infallibility of their Church. It is intended to bring back home those who have fallen away from their faith, to convince searching Christians of a home in the Catholic Church, and to convert in a gentle manner Protestants and Pseudo-Christians to the fullness of Christianity as found in the Catholic Church.

Ecce Fides is a work dedicated to defending Catholic beliefs through reason, Scripture, and the life of the Holy Spirit.

In the first three hundred years of Christianity, one million Catholics lost their lives as martyrs. Come join the faith that was built on Christ and the blood of his martyrs.

Ecce Fides: Pillar of Truth
by
Fr. John Pasquini
Shepherds of Christ Ministries

INTRODUCTION
by His Eminence Paul Cardinal Poupard

Of vital importance for evangelization is an awareness of the cultures in which people live, shape their thinking, develop their lifestyles, and seek to grow in their humanity. In our task of spreading the Good News we are called on to read the signs of the times and renew the way we participate in the Church's evangelizing mission, adapting to the new situations emerging in our continually changing cultures. But as cultures and currents of thought are fickle, changing under influences both good and bad, it is important for us to remain focused on the fullness of our faith if we are to carry out our mission as Christians, the great commandment to go out and preach the good news to all the peoples. Our faith, revealed in biblical culture, further understood through 2000 years of changing cultures, remains our response to the same Good News, the Gospel, the story of salvation in which God sent his Son to redeem the world through his birth, death, and resurrection. Through the Church, the bride of Christ, his light shines visibly to all men; by the power of the Spirit, in the sacraments instituted by Christ and entrusted to the Church, we can be healed and transformed, conformed to the same Son of God in preparation for eternal life. This is the truth we need to hear, meet, interiorize, and live out in conversion to God's own Divine will.

The Church is the custodian of the Good News and is also described by my patron, St Paul, in his letter to Timothy, as the pillar of truth, the subtitle of this useful catechetical book. The same Apostle to the gentiles took his missionary zeal from his awareness that ours is not a faith limited to its Hebrew roots, but contains within it a zeal to engage in a dynamic and salvific encounter with all cultures, for, as John Paul II would later proclaim magisterially, "a faith that does not become culture, is a faith which is not fully thought through, not truly lived out and not faithfully lived" (John Paul II, Letter creating the Pontifical Council for Culture, 20 May 1982). The Church is also expert in humanity, it safeguards the truth, not as a treasure hidden under a lampshade or in a bushel, but as an announcement *for* all humanity. Contrary to the characteristics of many sects and new alternative religions springing up today, the Church's essential truth is no secret, but a series of historical events that we need to proclaim in a way our contemporaries can access, *ad modum recipientis* as I was taught in seminary. We seek to hand this faith on to new generations with new energy, new ardour, and daring a new creativity in charity (cf. *Novo Millennio Ineunte*). And let us add, with a word characterizing Benedict XVI's Pontificate, joyfully. Following the paradigm of the Incarnation, wherein God became Man for our sake, the light of truth, takes on the form and fullness of cultures, not to glory in its sinful nature, but to transform them anew in order to reconcile men with each other, recreating them in the image and likeness of the new Adam. The meeting between the one faith and the many cultures is fundamentally at the service of this new Christian humanism, which is why "the split between the Gospel and culture is the drama of our times" as Pope Paul VI stated in the Apostolic Exhortation, *Evangelii Nuntiandi* 1975. Pope John Paul II also described how "the future of man depends on culture", which is not mere neutral context, but the battleground for the new evangelization. He set out our task thus: "In fact, we have to reach out to people where they are, with their worries and questions, to help them find the moral and spiritual landmarks they need to live lives worthy of our specific vocation, and to find in Christ's call the hope that does not disappoint (cf. *Rom* 5,5), as we follow the method used by the Apostle Paul at the Areopagus in Athens (*Acts* 17, 22-34)." (Discourse to the Plenary Assembly of the Pontifical Council for Culture, 16 March 2002).

In Europe we have noticed a resurgence in interest in religious matters in the public sphere, but sadly this has coincided with a spread of religious ignorance. This has led to a new thirst that is more spiritual than religious, by which I mean that people are seeking something beyond materialism, but without engaging with the reality of the Church, ecclesial traditions, and the historical events to which She witnesses, the very ways God has chosen to reveal Himself and transform us. This is also a time in which sociologists speak of a disaffection with institutions and the phenomenon of believing without belonging, the sort of insipid relativist belief that shirks the challenge of objective truth (see my work, *Where is your God?* Chicago 2004). I note too how many western cultures are invaded by consumerism and individualism, trends that through globalization are spreading throughout the world. So it is important that as Church, as family, sustaining each other, and, in particular, praying for each other as the Association *Shepherds of Christ* does, that we return to the source of life, our faith, which is usefully exposed in this volume, and take it out to our contemporaries, evangelizing them and their cultures and inculturating the Gospel, with the humility to remember that it is the Holy Spirit who is the principle agent of the new evangelization. Let all who thirst come to the water and find a water that is truly satisfying!

Paul Cardinal Poupard
President of the Pontifical Council for Culture

PONTIFICIUM CONSILIUM
DE CULTURA

I
HOLY SCRIPTURES
AND
TRADITION

Where did the Bible come from?

If you undermine the Catholic Church, you undermine the Bible!
Anonymous

The Christianity of history is not Protestantism.... To be deep in history is to cease to be a Protestant.
John Henry Cardinal Newman, Convert

I would not believe in the Gospel, had not the authority of the Catholic Church already moved me.
Augustine of Hippo
Contra epistolam Manichaei, 5, 6: PL 42, 176.

Did the Bible fall out of the sky? Certainly not! The Bible is the Word of God, but it is the Word of God because the Holy Spirit guided the Catholic Church in determining it to be such.

In the early Church there was no set Bible. In fact, there were many gospels and writings floating around claiming authenticity. There was the Gospel to the Ebionites (quoted by Epiphanus and Irenaeus), the Gospel to the Egyptians (referenced by Clement of Alexandria), the Gospel to the Hebrews (known by Papias, Hegesippus, and Eusebius, and quoted by Clement of Alexandria, Origen, Cyril and Jerome), the Gospel of the Nazaroeans (known by Hegesippus and Epiphanius, and known and preserved by Origen, Eusebius and Jerome), the Secret Gospel of Mark (quoted by Clement of Alexandria), the Gospel of Truth, the Gospel of Perfection, the Dialogue of the Redeemer, the Gospel of Peter (known by Eusebius), the Gospel of the Twelve Apostles, the Gospel of the Seventy, the Gospel of Philip, the Gospel of Matthias, the Gospel of Jude, the Gospel of Mary, the Gospel of Andrew, the Gospel of Barnabas, the Protoevangelium of James (Justin Martyr and Clement of Alexandria make mention of this source), the Infancy Gospel of Thomas, the Infancy Gospel of James (known by Jerome), the Apocryphon of James, the Apocalypse of Peter, etc. There were the Acts of Andrew, the Acts of John, the Acts of Thomas, and so forth.

It is important to recognize that more than 100 works of writing were being considered as part of what would come to be known as the New Testament.

Furthermore, many of the works we accept as part of the New Testament today were not fully accepted into the canon of the Bible until the fourth century—and not without great and fervent debate. For example, Eusebius, the greatest Church historian of his time, writing around the year 324 AD, points out that the epistles of James, Jude, 2 Peter, 2 and 3 John and the epistle to the Hebrews as well as the book of Revelation were still not accepted as part of the Bible. Amphilochius of Iconium (ca. 340-394) explains:

Now I am to read the books of the New Testament. Accept only four Evangelists, Matthew, then Mark, to which add Luke. Count John in times as fourth, but first in sublimity of teachings. Son of Thunder, rightly he is called, who loudly announced the Word of God. Accept from Luke a second book also, that of the Catholic Acts of the Apostles. Add to these that Vessel of Election, the Herald of the Gentiles, the Apostle Paul, writing wisely to the churches: One epistle to the Romans, to which must be added two to the Corinthians, and the one to the Galatians, and that to the Ephesians, after which there is one to the Philippians, then those written to the Colossians, to the Thessalonians two, two to Timothy, to Titus and to Philemon, one to each, and to the Hebrews one. Some call that to the Hebrews spurious, but not rightly do they say it; for the gift is genuine. What then is left? Of the Catholic Epistles some say seven need be accepted, others only three: One of James, one of Peter, one of John, or three of John and with them, two of Peter, the seventh that of Jude. The Apocalypse (Revelation) of John is also to be considered.. Some accept, but most will call it spurious (4).

It was already the fourth century and the structure of the Bible was still being debated.

Things even get more complicated. When we look at the modern day accepted canon of the Bible, particularly the New Testament, we notice something quite interesting. Open up to any good Protestant Bible, such as the RSV or the NRSV, or to any good Catholic Bible such as the NAB or the NJB and you will notice something that might be shocking to many who overlook the introductions to the various New Testament books. But when we look at them we notice the following: The Gospel of Matthew as we have it today seems, according to the scholars, not to have been written by the disciple of the Lord, but by a Greek speaking convert. The Gospel of John, Revelation and the three epistles of John which make up the Johannine corpus seem to be more the product of a Johannine community than the apostle John. In terms of St. Paul's writings, 2 Thessalonians, Colossians, and Ephesians seem to have been most likely written by another writer. One and 2 Timothy and Titus seem to have been written by a disciple of Paul and not by Paul himself. Hebrews, for a long time attributed to Paul, is now a work whose authorship is completely unknown. One and 2 Peter seem to have questions regarding Peter as the author. Likewise, the same problems occur with James and Jude. (It is no coincidence that most modern scholars, as well as the *Catechism of the Catholic Church*, refer to "sacred authors" when making reference to the authorship of the books of the Scriptures.)

The question must be asked: Why do Protestants not go back and look at the books that the Catholic Church rejected and also look at those books that the Catholic Church accepted into the Bible? This situation in terms of the formation of the canon of the New Testament has to be deeply troubling for a Protestant brother or sister. How can the Protestant know that the New Testament is the Word of God if scholars are capable of proving that the authorship of many of the books of the New Testament is questionable? Furthermore, how can a Protestant know that the Catholic Church did not overlook an authentic work of one of the apostles? Maybe there is a treasure waiting out there to be found? If I were a Protestant, I would be reexamining every book in existence claiming apostolic authenticity. Yet Protestants do not. Protestants accept the Catholic Church's Bible, the entire New Testament, on the authority of the Catholic Church. Given this, the question must be asked: If Protestants accept the Bible as the Catholic Church has produced it—under the power of the Holy Spirit—why do they not accept the authority of the Catholic Church in its interpretation of the Bible?

You may think things are getting out of control at this point. Well, just think about this. We have no original surviving manuscripts of any of the books of the New Testament. They are all copies! Furthermore, the copies we have are not all the same. In fact, no two copies are exactly alike! Some have estimated that there are as much as 200,000 variations within the various biblical texts.

In your free time feel free to examine the following texts and see how different they are from each other: *Codex Vaticanus, Codex Sinaiticus, Codex Alexandrinus, Codex Bezae, Codex Ephraemi Rescriptus, Codex Washingtonensis, Codex Koridethianus*. Which codex is the perfect text? Which minister is quoting the correct Scripture reference?

The early Church had no set Bible for the first four centuries. In fact, the first letters written in the Church can only be traced to the year 48 AD, some fifteen years after the resurrection. And the Gospel of John can only be traced to approximately the year 110 AD, some 10 years after the death of John.

At this point in the reading of this text, one might be terribly shocked by what you have read, but don't despair! This is where Sacred Tradition (the life of the Holy Spirit within the Church) comes in. It is the apostles and their successors, the bishops, who guided the Church in the ways of the faith. It is only through the guidance of the pope and the bishops in union with him that a Bible started to take shape (This should not be a surprise to us since even during the time of the apostles crucial questions of faith and morals were debated over and decided upon by councils of the Church [Acts 15:1-29]).

A list of what would become the Bible was approved by Pope Damasus I in 382 and reaffirmed by Pope Innocent I in 411. This list of the books of the Bible becomes approved at the Councils of Hippo (393 AD), Carthage III (397 AD) and Carthage IV (419 AD). And it is not till the Council of Trent in 1546 that the canon becomes completely closed.

The formation of the Bible over centuries should not be a source of concern for us since the "Chosen People" of the Old Testament lived without any written Scriptures for centuries. The Hebrew Scriptures were the product of the writing down of Sacred Tradition.

Protestants accept the Catholic New Testament in its entirety. Why? Because of the authority of the Catholic Church!

It is one of the ironies of history that the founder of Protestantism, Martin Luther, had to admit that it was the Catholic Church that gave us the Bible:

We are obliged to yield many things to the Catholics, that they possess the Word of God, which we received from them; furthermore, we would know nothing at all about the Bible if it were not for the Catholics (Luther, *Commentary on John*).

Even Luther recognized the Bible was not self-authenticating. Luther recognized the authority of the Catholic Church in determining the Word of God.

If Protestants accept this authority in regard to the Bible, why don't they trust its authority in all issues regarding faith and morals? For if the Church was infallible in the fourth century in putting the Bible together, why would it not be infallible throughout the succeeding generations?

Furthermore, does it not make more sense that the Church that put the Bible together in the first place would be the Church with the gift of interpreting it most accurately? How blessed we are to be Catholics!

Would Jesus leave us in confusion?

As we are too well aware of the Bible is interpreted in a variety of ways. Who has the right answer, the right interpretation? Would God leave us in confusion? When we take a look at Bible scholars from the best universities in the world, we see something very puzzling. No matter how well-learned these Bible scholars are they come up with disagreements on religious issues. It is not as if each well-educated scholar is seeing something that the other scholar is not seeing. Top-notch scholars know each other's arguments inside out, yet there is disagreement. Not only is there disagreement, but when we look at the arguments from one denomination of the Christian faith, we find that that denomination's arguments seem quite logical, and when we look at the arguments of another denomination's version of the faith, we find that their arguments seem just as logical. How can this be?

For one thing, the way we approach the Bible affects our interpretation of the Bible. We come to the Bible, as with everything else in life, with a certain predisposition. Our culture, religion, family background, etc., all affect how much weight we place on one argument or another. That is why there are so many disagreements from the best of scholars.

The question still remains, however, "Would God leave us in such confusion?"

Let us take a look at the most basic teachings that most Christians take for granted today. In the early Church, groups argued over whether Jesus was a man, an image, or a phantom; whether he was partially human, partially divine, fully human and/or fully divine. They argued whether there was a Trinity, whether there were three co-equal and co-eternal Persons in one God.

These above issues have all been acknowledged as resolved by mainline Christians. Yet who did the resolving? It is the Catholic Church that fought off all heresies and taught the truth that is so well appreciated by others today. It is the Catholic Church that taught that Jesus was fully human, fully divine, without confusion, division or separation. It is the Catholic Church that taught that the Three Persons of the Trinity are co-equal and co-eternal, without confusion, division or separation. If the Catholic Church was infallible in determining all these truths, why would it not be infallible throughout the succeeding generations?

In the early Church many small groups such as the Docetists, the Gnostics, the Montanists, the Marcionists, and so forth, all fought over the true meaning of the faith. An answer to this confusion had to be found; after all, a faith that is in confusion is not a faith at all.

The theologian Tertullian (ca. 155-240), building upon the great insights of Irenaeus (ca. 130-220) and the early Church, taught that the true faith was to be found in the writings of the apostolic writers (Sacred Scripture) and in the life of their communities (Sacred Tradition, the life of the Holy Spirit within the Church). In other words, the true teachings of Christ were to be found in the communities set up by the apostles and upon their authentic "memoirs," which would eventually be collected into what would become known as the New Testament.

Irenaeus, a disciple of Polycarp, who in turn was a disciple of the apostle John and "a companion of the apostles," and appointed bishop of Smyrna by the apostle John, writes some very pertinent words when he describes in his tract *Against Heresies* (3,4,1) the following:

> *It is necessary to seek among others the truth which is easily obtained from the Church.... If there should be a dispute over some kind of question, ought we not have recourse to the most ancient churches [i.e., communities, dioceses] in which the apostles were familiar, and draw from them what is clear and certain in regard to that question? What if the apostles had not in fact left writings to*

us? Would it not be necessary to follow the order of Tradition, which was handed down to those to whom they entrusted the churches (The Faith of the Early Fathers, vol. 1, trans. William Jurgens, Collegeville: The Liturgical Press, 1970).

Tertullian, in *The Demurrer Against Heretics* (21:4; 7) (ca. 200) argued:

All doctrine which agrees with the apostolic churches [i.e., communities], those nurseries and original depositories of the Faith, must be regarded as truth, as undoubtedly constituting what the churches received from the apostles, what the apostles received from Christ, and what Christ received from God.... We communicate with the apostolic churches because there is no diversity of doctrine: this is the witness of the truth (Ibid.).

In section 32:1 we read:

Moreover, if there be any [heresies] bold enough to plant themselves in the midst of the apostolic age, so that they might seem to have been handed down by the apostles because they were from the time of the apostles, we can say to them: let them show the origins of their churches, let them unroll the order of their bishops, running down in succession from the beginning, so that their first bishop shall have for author and predecessor some one of the apostles (Ibid.).

It is quite clear that the true faith is to be found in the "memoirs" of the apostles and the Tradition of the Church. Those communities founded upon the apostles and their successors are the places where the true and authentic deposit of the faith can be found.

The Bible alone is insufficient and unchristian

The teaching of the Church has been handed down through an order of succession from the apostles, and remains in the churches even until the present day. That alone is to be believed which is not at variance with the ecclesiastical and apostolic tradition.
Origen (ca. 185-253)

Irenaeus reminds the faithful that the faith was brought into the barbarian lands not by the Bible, but by Tradition: "The Barbarian tribes received the faith without letters."

The need for Tradition along with the Scriptures is unquestionable. The early Church testifies to this. The very reality that the Scriptures flowed out from Tradition points to this reality.

The Bible alone is inadequate! The "Bible only" approach as a rule of faith is nowhere to be found in the Bible. In fact, the "Chosen People" of the Old Testament lived without the Hebrew Scriptures for centuries. The "Bible only" approach was an invention of primarily the 14th century heretic John Wycliffe—a prototype of Protestantism—and the sixteenth century theologian Martin Luther, the first Protestant. It was an invention that was radically contrary to the history and nature of Christianity and Judaism.

God speaks to us in the Bible but also beyond the scope of the Bible. God speaks to us by means of natural revelation (Rom. 1:20; Wis. 13:1-9), by means of Jesus' life and words (Jn. 1:1,14; Lk. 4:44; 5:1), by inspiration (Lk. 3:2-3; Acts 4:31; Heb. 4:12-13), and by the oral preaching of the Gospel (1 Thess. 2:13).

Sacred Tradition is essential. As Papias (ca. 67-130), the bishop of Hierapolis in Asia Minor, "a hearer of St. John" the apostle and a friend of Polycarp—the disciple of St. John—(*Against Heresies*, 5, 33) as well as an "acquaintance" of the other apostles, and a friend to the daughters of the apostle Philip (*Ecclesiastical History*, 3, 39), states in his *Explanations of the Sayings of the Lord*:

> *I shall not hesitate to set before you, along with my own interpretation, everything I carefully learned from the elders and carefully remembered.... It seemed to me that I could profit more from the living voice [of Tradition] than from books.*

The Bible itself points to the need for Tradition: Luke reminds us that his Gospel is the writing down of what has been handed down to him (Lk. 1:1-4). The apostle John reminds us that there are not enough books in the world to describe what Jesus did (Jn. 20:30; 21:25) and that often when he communicated with his own disciples he did not use pen or ink, but spoke face to face (2 Jn. 1:12; 3 Jn. 13). Paul, Timothy, and Jude remind us strongly to hold firm to the traditions that have been handed down by word of mouth and by letter (1 Thess. 2:13; 2 Thess. 2:15; 1 Cor. 11:2; 2 Tim. 1:13; 3:14; Jude 17).

When we as Catholics speak of Sacred Tradition we are not talking about human traditions, such as that which is alluded to in Matthew 15:3, 6-9 or Colossians 2:8, rather what we are pointing to is tradition with a big T; Traditions that were handed down in the Church by Jesus and his apostles (Lk. 1:1-4; 10:16; Jn. 21:25; Acts 2:42; 1 Cor. 15:3, 11; 2 Thess. 2:15; 2 Tim. 2:2).

As Athanasius (360 AD) wrote in *Four Letters to Serapion of Thimius* (1:28):

> *Let us note that the very tradition, teaching, and faith of the Catholic*

Church from the beginning, which the Lord gave, was preached by the apostles, and was preserved by the Fathers. On this was the Church founded; and if anyone departs from this, he neither is nor any longer ought to be called a Christian....

Or as Origen (ca. 230) wrote in *Fundamental Doctrines* (1, Preface 2):

The teaching of the Church has indeed been handed down through an order of succession from the apostles, and remains in the churches [dioceses] even to the present time. That alone is believed as the truth which is in no way at variance with ecclesiastical and apostolic tradition.

God has not left us in confusion. He has given us Sacred Tradition and Sacred Scripture. And in the event that confusion should still remain, he left us a teaching office.

A teaching office—what we call the Magisterium—is needed in the Church to determine the true interpretation of the faith, since private interpretation can often lead to heresy. As 2 Peter 3:16 states: There are "certain things hard to understand, which the unlearned and unstable distort, as they do also the other scriptures, to their own destruction" (see also 2 Pet. 1:20). In Acts 8:26-40 we read:

Then the angel of the Lord spoke to Philip, "Get up and head south on the road that goes down from Jerusalem to Gaza, the desert route." So he got up and set out. Now there was an Ethiopian eunuch, a court official of the Candace, that is, the queen of the Ethiopians, in charge of her entire treasury, who had come to Jerusalem to worship, and was returning home. Seated in his chariot, he was reading the prophet Isaiah. The Spirit said to Philip, "Go and join up with that chariot." Philip ran up and heard him reading Isaiah the prophet and said, "Do you understand what you are reading?" He replied, "How can I, unless someone instructs me?" So he invited Philip to get in and sit with him.

This was the scripture passage he was reading: "Like a sheep he was led to the slaughter, and as a lamb before its shearer is silent, so he opened not his mouth. In (his) humiliation justice was denied him. Who will tell of his posterity? For his life is taken from the earth." Then the eunuch said to Philip in reply, "I beg you, about whom is the prophet saying this? About himself, or about someone else?" Then Philip opened his mouth and, beginning with this scripture passage, he proclaimed Jesus to him. As they traveled along the road they came to some water, and the eunuch said,

"Look, there is water. What is to prevent my being baptized?" Then he ordered the chariot to stop, and Philip and the eunuch both went down into the water, and he baptized him. When they came out of the water, the Spirit of the Lord snatched Philip away, and the eunuch saw him no more, but continued on his way rejoicing. Philip came to Azotus, and went about proclaiming the good news to all the towns until he reached Caesarea.

Philip the deacon, the representative of the Church, was needed to give the true meaning of the Scriptures to the eunuch.

We see this same pattern in Acts 15 at the Council of Jerusalem. A conflict arose in the early Church (ca. 50 AD) around what to do with Gentile converts. Should they first convert to Judaism through circumcision and then follow the Jewish dietary laws or should they be admitted into the Church without the need to follow these regulations.

Paul was the leader of the opposition who believed that there was no need for circumcision and the dietary laws under the new law of Christ. In 49 AD Paul and some of his associates journeyed to Jerusalem to confer with the apostles, and in particular with the head of the apostles, Peter. After much discussion, Peter and James ruled that Gentile converts would not be required to observe the Jewish regulations.

Again we see the need for a Magisterium. Even Paul recognized the need for seeking the Church's advice and approval for the correct interpretation of the Scriptures and God's will.

Originally the teaching office was made up of the apostles, the first bishops, and Peter, the first pope (Acts 15:1-35). With every succeeding generation the successors of Peter and the other apostles were given charge of protecting the faith from errors—with Peter and his successors having primacy of power (Mt. 16:18f).

We are reminded of the importance of listening to the Church, the Pillar of Truth (1 Tim. 3:14-15; Mt. 18:17-18; Lk. 10:16) in its authority to teach (Mt. 28:20) to interpret the Scriptures (Acts 2:14-36; 2 Pet. 1:20-21; 2:1; 3:15-17) and to bind and loose (Mt. 18:18; Acts 15:28-29).

Sacred Tradition, Sacred Scripture, and the Magisterium are inseparable realities. All three are necessary to assure the proper transmission of the faith.

Vincent of Lerins (ca. 450) in his *Commonitoria* (2,1-3) beautifully illustrates the need for Sacred Tradition, Sacred Scripture, and the teaching office of the Church (the Magisterium) when seeking the authentic word of God.

With great zeal and closest attention...I frequently inquired of many men eminent for their holiness and doctrine, how I might, in a concise and, so to speak, general and ordinary way, distinguish the truth of the Catholic faith from the falsehood of heretical depravity. I received almost always the same answer from all of them, that if I or anyone else wanted to expose the frauds and escape the snares of the heretics who rise up, and to remain intact and sound in a sound faith, it would be necessary, with the help of the Lord to fortify that faith in a [pertinent] manner: first, of course, by the authority of the divine law; and then, by the Tradition of the Catholic Church. Here, perhaps, someone may ask: 'If the canon of the Scriptures be perfect, and in itself more than suffices for everything, why is it necessary that the authority of ecclesiastical interpretation be joined to it?' Because, quite plainly, Sacred Scripture, by reason of its own depth, is not accepted by everyone as having one and the same meaning. The same passage is interpreted by others so that it can almost appear as if there are as many opinions as there are men. Novatian explains a passage in one way, Sabellius another, Donatus in another; Arius, Eunomius, Macedonius in another; Photinus, Apollinaris, Priscillian in another; Jovinian, Pelagius, Caelestius in another.... [Without reference to the Tradition as expounded and taught by the apostles and their successors, the bishops, there would be no way of knowing the true meaning of the Scriptures.] (Jurgens, vol. 3).

The Bible "only" approach to divine revelation is unbiblical and contrary to Sacred Tradition—which we are commanded to hold onto (2 Thess. 2:14-15). The Bible "only" approach is a "human" tradition or invention which is contrary to the deposit of the faith (Matt. 15:3, 6-9; Col. 2:8).

We must remember that the Bible tells us that it is the Church and not the Bible that is the "pillar and bulwark of the truth" (1 Tim. 3:15). We must remember what the founder of Protestantism, Martin Luther (1517) had to acknowledge:

We are obliged to yield many things to the Papists [Catholics]—that they possess the Word of God which we received from them, otherwise we should have known nothing at all about it (Commentary on St. John, 16).

What Protestants can't answer!

How can we know with certainty what belongs in the Bible? The Book of Mormon, the Quran, the writings of the Hindus and of the Buddhists, the writings of Mary Eddy Baker, and all the books that Christianity would exclude from the Bible all claim to be self-authenticating! Philosophically speaking and hermeneutically speaking, documents cannot authenticate themselves! An outside source is always needed to authenticate a document.

So it is with the Bible. An infallible Church, the Catholic Church, founded by Jesus Christ authenticated the books that would make up the Bible. The Bible alone approach is unchristian and intellectually unsupportable! Again, I repeat the words of Martin Luther:

> *We are obliged to yield many things to the Papists [Catholics]—that they possess the Word of God which we received from them, otherwise we should have known nothing at all about it (Commentary on St. John, 16).*
>
> *Martin Luther, Founder of Protestantism*

Two forms of Revelation

What many Protestants fail to realize is that there are two forms of God's revelation. One form is often referred to as natural revelation and the other as divine revelation. Natural revelation is divided into two forms, the revealing of God's presence through the material universe (Rom. 1:20; Wis. 13:15) and the revelation of God through the natural and moral law that is embedded in the core of all human beings (Rom. 2:15)—this law becomes perceptible by a clear and informed conscience. Finally, there is divine revelation or the deposit of the faith, as found in Sacred Tradition—the life of the Holy Spirit working through the life of the people of God—and Sacred Scripture—which flowed out of Sacred Tradition. Divine revelation helps to correct the misinterpretations of natural revelation. When we seek truth and God we must seek to grasp the entirety of God's revelation to us!

What about Revelation 22:18-19?

Revelation 22: 18-19 states: "I warn everyone who hears the words of prophecy of this book: if anyone adds to them, God will add to that person the plagues described in this book; if anyone takes away from the words of the book of this prophecy, God will take away that person's share in the tree of life and in the holy city, which are described in this book."

Many argue that by resorting to "tradition" one is contradicting Revelation 22. This is not the case. Tradition does not add as much as clarifies the word of God in the Scriptures. The life of the Spirit in the Church (Sacred Tradition) helps us to understand how the Church always understood a particular scripture passage.

For example, for Catholics, John 6:35-71 testifies to the real presence of Christ in the Eucharist. Yet many do not accept this Catholic and Orthodox understanding of the Eucharist. So as Catholics we look to Sacred Tradition: Ignatius of Antioch, a disciple and friend of the apostles John, Peter, and Paul wrote in 107, only 7 years after the death of the apostle John, that "only heretics abstain from the Eucharist...because they do not confess that the

Eucharist is the Flesh of our Savior Jesus Christ." The Church has always maintained the real presence of Christ in the Eucharist. It is only in the 16th century, with the birth of Protestantism, that Protestants began to slowly reject this reality.

Secondly, the Bible is full of additions. The most legendary is the ending of Mark's Gospel. There are three different endings to Mark's Gospel that can be found in ancient manuscripts. For example, Jerome quotes a fourth century manuscript known as the "Freer Logion" (which appears after v. 14). It is usually footnoted in Bibles. But all Bibles contain within the main text a "longer ending" (16: 9-20) and a "shorter ending" (found after v. 8 or v. 20) to the Gospel of Mark.

If you have two different endings in the same text, you clearly have additions. All scholars acknowledge that the "longer ending" is not from Mark and that it was likely written during the second century. Furthermore, the "shorter ending" appears mainly in seventh to ninth century manuscripts of the Bible as well as in an old Latin version of the Bible. Yet both endings are in the Bible and are considered the Word of God, Sacred Scripture. If you check the footnotes in your Bible you will see this explained.

Likewise, when we look at the book of Genesis we notice that Chapter 1 to Chapter 2:1-4 has one account of the creation of the world and Chapter 2:4f has a second account. That is why the Bible has as the heading to Chapter one, "The First Story of Creation," and as a heading to Chapter two, "The Second Story of Creation." If you have two different accounts in the same text, you clearly have additions.

Thirdly, we know that some subtractions have been made since some parts of the Bible are lacking. Let us take just one book of the Old Testament to make our point, the book of Job: Where are the words to Job 24:19-21? Where are the words to Job 28:3-4? Where is verse 30 of Job 34? What happened to Chapter 36, verses 16-20? Clearly, some things have been lost or subtracted or so obscured that they are not represented in the Bible!

What serves to compensate for this? Sacred Tradition! The successors of the apostles, the bishops, express the Spirit in the life of the Church and assure that nothing is lacking in divine revelation.

Fourthly, the book of Revelation itself, like all the books of the Bible, has ambiguities and additions or subtractions in it. For example, the expression "freed us" in Revelation 1:5 is translated in some ancient manuscripts as "washed us"; in 9:13 "four horns" is often found in manuscripts as "horns"; in 11:12 "they" is often found in manuscripts as "I"; in 12:18 "then the dragon took" is often found in manuscripts as "then I stood"; in 13:18 "666" is often found in manuscripts as "616"; in 15:3 "key of the nations" is often found in manuscripts as "king of the ages"; in 15:6 "bright linen" is often found in manuscripts as "stone"; in 18:2 "hateful beast" is often found in manuscripts as "hateful bird"; in 18:3 "for all nations have drunk" is often found in manuscripts as "she has made all nations drink"; in 19:13 "dipped in" is often found in manuscripts as "sprinkled with"; in 22:14 "washed their robes" is often found in manuscripts as "do his commandment"; in 22:21 "the grace of the Lord Jesus be with all the saints" is often found in manuscripts as "the grace of the Lord Jesus be with you." And when you try to find, in some ancient manuscripts, Revelation 13:7, you will not find it.

As was mentioned earlier, the Bible has some 200,000 variations in the manuscript evidence.

If we are to take **Revelation 22:18-19** in such a literalistic view, then how do we explain Deuteronomy 4:2: "You must neither add anything to what I command you nor take away anything from it..." If Deuteronomy 4:2 is taken in such a narrow way, then we as people of God should reject Revelation 22 and the whole New Testament since it came after Deuteronomy 4:2. Obviously this is not what was meant by Deuteronomy or Revelation.

Finally, Protestants violate their own doctrine. In the 16th century they eliminated seven entire books from the Old Testament and parts of two others (Tobit, Judith, 1 and 2 Maccabees, Wisdom, Sirach, Baruch and parts of Daniel and Esther).

The New Testament quotes the Old Testament approximately 350 times, and in approximately 300 of those instances (86% of those instances), the quotation is taken from the Septuagint version of the Old Testament which contains the books that Protestants eliminated. Also, the deuterocanonical books that the Protestants eliminated are quoted in the New Testament not less than 150 times. Protestants violate their own principle.

The meaning of the text is that the Word of God must be accepted authentically and completely. To do so means that one must look to the life of the Holy Spirit within the Church (Sacred Tradition) and the written inspired words of God (Sacred Scripture). And finally, the successors to the apostles in union with the head of the apostles, Peter, the bishops and popes, are entrusted with interpreting this Sacred Scripture and Sacred Tradition.

Is all Scripture to be interpreted in the same way?

As Catholics we seek to understand the Scriptures in the way they were meant to be understood. We allow the Scriptures to say what they want to say (*Exegesis*) as opposed to making them say what we want them to say (*Eisegesis*).

Many people take a particular belief system and then go to the Scriptures and try to find justification for their belief system by forcing a completely foreign interpretation into a particular Scripture passage.

We should never fear the Lord. Let him speak to us the way he intended.

As Catholics we seek to comprehend the *intent* of the inspired authors in their writings. We also try to comprehend the various *senses* in which the Scripture passages were written. Finally, we seek to understand the Scriptures in light of the same *Spirit* in which the inspired writers wrote them.

The Intent

The following provides a helpful guideline regarding author intent:

1) What condition was the author confronting?
2) What was the culture of the area like?
3) What literary genres were common at the time?
4) What modes of feeling, narrating, and speaking were common at the time?

For example, the book of Revelation addresses a Church under persecution by either Nero or Domitian. The sacred author is seeking to encourage the faithful to persevere in Christ amidst great trials and tribulations. "Hold on," "stand fast," victory is at hand for those who remain loyal to God.

The author uses symbolic and allegorical language characteristic of apocalyptic or resistance literature. Apocalyptic literature makes use of visions, animals, numbers, and cosmic catastrophes in a coded language with the express purpose of instructing the faithful in times of difficulty. The very nature of apocalyptic literature—which enjoyed great popularity amongst the Jews and Christians during the first two centuries—was ideal for conveying a secret message to Christians that could not be readily understood by the enemies of Christianity.

The Senses

In terms of the senses of Scripture, the following are important to keep in mind.

1) What is the literal meaning of the text?
2) What is the spiritual sense of the text?
3) What is the allegorical sense?
4) What is the moral sense?
5) What is the anagogical sense?

In the "passion and resurrection narratives" (Mt. 26f; Mk. 14f; Lk. 22f; Jn. 18f) we have the literal reality that Jesus Christ suffered, died, and rose from the dead.

In terms of the spiritual sense of these narratives we recognize that Christ's death and resurrection was for our salvation—that in Christ we are born to eternal life. We also recognize, amongst other insights, that Christ's death made all suffering redemptive.

In terms of the allegorical sense, Jesus can be seen as the "New Moses." Moses freed the people of God from slavery and brought them to the edge of the "promised land" "flowing with milk and honey." In a much more powerful manner, Jesus, as the new and greater Moses type or figure, freed us from the slavery of sin and brings us into the eternal bliss of heaven. Likewise, the crossing of the Red Sea by Moses is seen as being symbolic of baptism as well as a sign or type of Christ's victory over death. As for the moral sense that can be acquired through a reading of these narratives, the insights are unending. The moral sense is intended for, as Paul states, "our moral instruction" (1 Cor. 10:11). Jesus reminds us that being moral entails the seeking and fulfilling of the will of the Father (cf. Mt. 26:39).

The anagogical sense of the passion and resurrection narratives focus on realities and events in terms of their eternal significance. The resurrection of Jesus is a sign to us that we too, in him, will likewise rise and be brought into eternal glory after the end of our earthly journey. The anagogical sense is intended to guide us toward eternal life with God in heaven.

The four above senses are beautifully summarized by a medieval couplet that states: "The Letter speaks of deeds; Allegory to faith; The Moral how to act; Anagogy our destiny."

The Spirit

In terms of interpreting the Bible in light of the Spirit in which it was written we pursue the following rules:

1) How is a particular Scripture passage understood within the context of the whole Bible?

2) How is the Bible understood within the Tradition it came out from? If an interpretation of a particular passage makes a person conclude that Jesus was only a phantom or spirit, then one cannot accept this as being an authentic Tradition of the Church. One must reject this interpretation as not being faithful to the life of the Holy Spirit within the Church.

3) How is the passage of the Bible understood in terms of a coherence of truths? All the doctrines of the Church must fit together like a puzzle. You cannot have one belief contradicting another belief. There can only be one coherent truth.

Each Scripture passage is like a piece of a puzzle that depicts a picture. One piece of the puzzle is insufficient for understanding and recognizing what is being portrayed by the whole of the puzzle. One needs all the pieces, or at the very least, the core pieces. The same can be said of the Scriptures. One passage needs to be understood within the context of the whole of the Scriptures for a true interpretation of the Word of God.

This prevents the use of a technique used by many fundamentalists and pseudo-Christians called "proof-text" theology. Let me use an example. In one passage of the Bible Jesus says "Blessed are the peacemakers" (Mt. 5:9), yet in another he says, "I have not come to bring peace, but division" (Lk. 12:51). This may seem a contradiction but it is not. Jesus is pointing out that in the process of being a peacemaker one will inevitably come up against obstacles which could inevitably lead to division. When the one text is understood in terms of the other, both make perfect sense in terms of the Christian way of life. But if one passage is taken without the other passage, confusion can occur regarding Jesus' true teaching.

Another example of the need to interpret a particular Scripture passage within a coherent and historically accurate context within the whole of the Bible is seen in Jesus' words on the cross: "My God, my God, why have you forsaken me" (Mt. 27:46). At first glance this makes Jesus appear as a man on the verge of despair. Yet when this passage is taken within the context of the whole of the Scriptures we see that the contrary is true. Far from being an echo of despair, the words "My God, my God, why have you forsaken me" are an affirmation of Jesus' identity as the Savior and Messiah. "My God, my God, why have you forsaken me" are the first words of Psalm 22 of the Old Testament which foretell of the passion and triumph of the Messiah (While it is true that Jesus suffered the pangs of abandonment, he did not despair; furthermore, the words Jesus proclaimed were far more profound than the pangs of suffering.). If I were to say "Our Father, who art in heaven" everyone would know that I was reciting the first words of the "Our Father"; likewise when Jesus said "My God, my God, why have you forsaken me" the Jewish people would have been fully aware that Jesus was making reference to Psalm 22. He was reminding them that he was the fulfillment of Psalm 22—that he was the Savior and Messiah! (Observe the stunning similarities that exist between Psalm 22 and the passion narratives in the Gospels).

If one took Matthew 27:46 out of context, one could end up with a distorted vision of Jesus' human and divine natures.

Another example comes from the Old Testament vision of God as the "Warrior God." Many people in today's culture find this absolutely abhorrent and refuse to accept what the Scriptures make very clear. These people find the image of a "Warrior God" as abhorrent because they fail to recognize the proper context of the Old Testament image.

In the ancient world, war was part of everyday life. In fact, kings often waited for the good weather of spring to begin new campaigns of war: "the kings go out to war…in the spring of the year" (2 Sm. 11:1). In a world where war is the norm, it does not seem abhorrent to view God as a "Warrior God." For the Jews, God would be there to save them from the attacks of their enemies if they remained faithful to the laws and commandments of Moses, but if they did not, they would be chastised by God by means of defeat. The defeat by the Assyrians and the Babylonians was a mark, according to the Old Testament prophets, of God's chastisement for failing to be faithful to the covenant made between God and his people.

The most egregious abuse of this principle of interpretation today is found amongst fundamentalists who proclaim the "gospel of wealth." This is quite common on fundamentalist Christian television. The argument goes: If you are faithful to God, he will grant you a long life, no suffering, and great wealth. If you are unfaithful you will end up having a short life with great suffering, and you will die a pauper.

This vision of life is strongly emphasized in the early books of the Hebrew Scriptures (the Old Testament), particularly in the wisdom literature, and is often referred to as the "theory of retribution." The key point of this theory is that an ordered and moral society is one that fosters justice and harmony.

Having said this, however, if these Scripture passages were all we had, we would have a distorted image of the Word of God. The sad reality is that so much of modern Christianity focuses on the "theory of retribution" that the Word of God inevitably becomes unrecognizable.

The book of Job is essential for putting a proper perspective on the earlier writings of the Scriptures. The book of Job is a turning point in the Old Testament; it is a key part of the puzzle that puts all into focus.

In the book of Job we are taught that one's finite mind cannot grasp the mysterious plan of God. God's ways are not our ways. Our call is to remain faithful, and in doing so, we help to fulfill God's providential plan for ourselves and the world. This is beautifully illustrated in the words of an unknown civil war soldier:

I asked for strength that I might achieve; I was made weak that I might learn humbly to obey. I asked for health that I might do greater things; I was given infirmity that I might do better things. I asked for riches that I might be happy; I was given poverty that I might be wise. I asked for power that I might have the praise of men; I was given weakness that I might feel the need of God. I asked for all things that I might enjoy life; I was given life that I might enjoy all things. I got nothing that I asked for, but everything that I had hoped for. Almost despite myself, my unspoken prayers were answered; I am, among all men, most richly blessed.

It is for this reason that even the just suffer. We are all aware of the many good people who have died young and poor, and let us never forget that our Savior was poor, young, and suffered death on the cross for us.

The book of Job adds one more piece to the puzzle regarding our understanding of God's mysterious ways!

Seeking to understand the Scriptures the way they were meant to be understood is at the heart of the Catholic Church's approach to the Word of God. We as Catholics seek to understand the intent, the senses, and the spirit of the Scriptures.

Why was the Catholic Church careful in making Bibles available to individual believers?

First, the Church knew that the Bible in the hands of the untrained would lead to disunity and heresies. History has proven this. There are currently some 33,000 Protestant denominations and 150,000 pseudo-Protestant denominations. Without an authoritative teaching office—the bishops in union with the pope—disunity and heresies cannot help but swell in numbers.

Second, the populations of the world prior to modern times were mostly illiterate. So even if people had a Bible in their hands they would not have been able to read it. Preaching and the use of art and stained-glass windows were for the vast majority of people the only means of learning about the Gospel message.

Third, for those people who could read and wanted a Bible, the cost of Bibles was exorbitant. Bibles prior to the invention of the printing press in the 15th century were hand written by monks and took years to produce. This was costly and made the purchase of Bibles impossible for the vast majority.

Finally, the assertion that Catholics never read the Bible is absurd. We wrote it; we put it into a canon; and we developed our theology from it. Any reader of the ancient Catholic writers from the 1st century to the present can see this!

The Bible is at the core of the Church's liturgy. Catholics who go to Mass every day of the week will essentially hear the entire Bible read in two years. For those who attend Mass on Sundays only, they essentially hear the entire Bible read in a three year period. Furthermore, the entire canon of the Mass is Scriptural, from the "Greeting" to the final "Dismissal."

How many denominations can say this?

Why do Catholics have more books in the Old Testament than Protestants or Jews?

During the Protestant Reformation Martin Luther (ca. 1534), after losing a debate against the great Catholic scholar Johann Eck on the topic of purgatory, decided to drop seven books from the Old Testament—many of which Johann Eck made reference to in defense of the Catholic faith—1 and 2 Maccabees, Sirach, Wisdom, Baruch, Tobit, Judith and parts of Daniel and Esther (These books are often referred to as deuterocanonical books). Martin Luther only made this momentous decision some seventeen years after his founding of the Protestant movement. Why did he not make these changes in 1517? What made him change his mind all these years later? The answer is that these books were a problem for his theology and the theology of Protestantism and thus had to be eliminated.

Despite Luther's predicament with Johann Eck, another reason for dropping these books from the canon of the Scriptures by Protestants was because they were written in Greek and found in the Septuagint (Greek) version of the Hebrew Scriptures. It is for this same reason that the Jewish people in the year 90 to 100 AD excluded these books. Many today argue that since these books are not in Hebrew and since the Jewish people today do not have these books in their Hebrew Scriptures, then they do not belong in the Bible.

As Catholics we would respond by taking a closer look at history before coming to such a quick conclusion. The first thing to recognize is that at the time of Jesus these deuterocanonical books were accepted as Scripture. Furthermore, the New Testament quotes the Old Testament approximately 350 times, and in approximately 300 of those instances (86% of those instances), the quotation is taken from the Greek, not Hebrew, Septuagint version of the Old Testament which contains the books that Protestants and Jews eliminated. It is only after the Jewish Council of Jamnia (ca. 90-100), after the fall of Jerusalem (ca. 70), and after an official break between the Pharisees and Jewish Christians, that a change occurs.

The Pharisees recognized that more and more Christians were coming from the Greek speaking Gentile world, and in order to distinguish themselves from the Christians they sought to remove all traces of Greek from their Scriptures. (Ironically, a lot of the Greek versions of the books they took out have in recent years been found in the original Hebrew).

It is crucial for us as Christians to recognize that the Greek Septuagint (LXX) version of the Scriptures was used by Jews throughout the Greek speaking world and was recognized as inspired prior to the Jewish Council of Jamnia (ca. 90-100).

Another important point is that by the time of the Jewish Council of Jamnia (ca. 90-100) the Christian Church (ca. 33) had already been established as the authoritative determiner on all matters concerning faith and morals, which included the formation of the canon of the Scriptures. Furthermore, the Jewish Council of Jamnia never made a statement regarding the closing of the Canon.

It is important to reiterate that the early Church always accepted the deuterocanonical books as part of the Scriptures. It was often quoted in the early Church (i.e., the *Didache* 4:5 (ca. 70); *Barnabas* 6:7 (ca. 74); *Clement* 27:5 (ca. 80), etc.).

However, the most important reason why the deuterocanonical books were accepted by the Catholic Church is because the apostles themselves accepted them. The apostles often quoted from the Greek Septuagint version of the Scriptures, thereby affirming its importance and validity. For example, compare Matthew 1:23 with Isaiah 7:14. Matthew is quoting from the Septuagint version of the Scriptures, the same version that holds the deuterocanonicals. Another example can be found in Luke's Gospel. Luke chapter 1:5 to chapter 3 is entirely constructed from the Septuagint version of the Bible.

As you look throughout the New Testament footnotes you will find the abbreviation for the Septuagint, LXX, throughout. There are 340 places where the New Testament quotes the Septuagint and only 33 places where the Hebrew only version of the Bible is quoted.

The point is that if the Greek Septuagint was good enough for the apostles, it is good enough for us Catholics.

Let us look at the following passage from the deuterocanonical book of Wisdom:

Let us lay traps for the upright man, since he annoys us and opposes our way of life, reproaches us for our sins against the Law, and accuses us of sins against our upbringing. He claims to have knowledge of God and calls himself a child of the Lord. We see him as a reproof to our way of thinking, the very sight of him weighs our spirits down; for his kind of life is not like other people's and his ways are quite different. In his opinion we are counterfeit; he avoids our ways as he would filth; he proclaims the final end of the upright as blessed and boasts of having God for his father. Let us see if what he says is true, and test him to see what sort of end he will have. For if the upright man is God's son, God will help him and rescue him from the clutches of his enemies. Let us test him with cruelty and with torture, and thus explore this gentleness of his and put his patience to the test. Let us condemn him to a shameful death since God will rescue him....(Wisdom 2:12-20, NJB).

This passage was written approximately one century before the crucifixion of Christ, yet one cannot but be amazed at the similarity between this passage and the passage describing the Passion of our Lord and Savior. We have here in the book of Wisdom the pre-figuration of the crucifixion of Jesus Christ.

It is not surprising that the Protestant Reformers never completely threw out the deuterocanonical books of the Old Testament. They saw them as worthy of being kept in an appendix. To this very day, these books are found in an appendix to the Old Testament. This very act is a testament to the discomfort that these sixteenth century revolutionaries had and modern day Protestants have in eliminating the deuterocanonical books.

II
THE CHURCH

Who's your founder?

The blessed apostle Paul teaches us that the Church is one, for it has 'one body, one spirit, one hope, one faith, one baptism, and one God.' Furthermore, it is on Peter that Jesus built his Church, and to him he gives the command to feed the sheep; and although he assigns like power to all the apostles, yet he founded a single chair, and he established by his own authority a source and an intrinsic reason for that unity. Indeed, the others were that also which Peter was; but a primacy is given to Peter, whereby it is made clear that there is but one Church and one Chair—the Chair of Peter. So too, all are shepherds, and the flock is shown to be one, fed by all the apostles in single-minded accord. If someone does not hold fast to this unity of Peter, can he imagine that he still holds the faith? If he deserts the chair of Peter upon whom the Church was built, can he still be confident that he is in the Church?

Cyprian of Carthage (ca. 251)
De Catholicae Ecclesiae Unitate, 2-7

If we want to find the true Christian faith—in all its fullness—we need to look at its foundation. Depending on whatever statistics we look at there are anywhere from 33,000 to 150,000 groups, cults, and denominations each claiming to have the authentic Christian faith.

Who is right? By looking at the founders of these groups we can come up with some key insights. For the purpose of this work, we will look at the founders of the main Christian and pseudo-Christian ecclesiastical communities in the United States and Europe.

All quality historians, from Harvard to Oxford, and all quality history books, whether Catholic or secular, recognize Jesus as founding the Catholic Church (ca. 33 AD). More will be said about this later.

Now let us look at some of the Protestant and pseudo-Christian ecclesiastical communities. Remember, there was no such thing as a Protestant Church until the sixteenth century; Jesus can never be claimed as the founder of any Protestant denomination. Let us look at some of their founders:

Denomination	Founder
Lutherans	Martin Luther (ca. 1517)
Anabaptists	Nicholas Storch / Thomas Munzer (ca. 1521)

Denomination	Founder
Swiss Reformed	Ulrich Zwingli (ca. 1522)
Hutterites	Jacob Hutter (ca. 1528)
Anglicans	Henry VIII (ca. 1534)
Calvinists	John Calvin (ca. 1536)
Familists	Hendrik Niclaes (ca. 1540)
Unitarians	Michael Servetus (ca. 1553)
Presbyterians	Calvin/ John Knox (ca. 1560)
Arminianism	Jacobus Arminius (ca. 1560-1609)
Puritans	T. Cartwright (ca. 1570)
Congregationalists	Robert Brown (ca. 1582)
Baptists	John Smyth (ca. 1609)
Dutch Reformed	Michaelis Jones (ca. 1628)
Quakers	George Fox (ca. 1650)
Mennonites	Menno Simons (ca. 1653)
Cameronians	Richard Cameron (ca. 1681)
Pietism	Philip Jacob Spener (1675)
Amish	Jakob Amman (ca. 1693)
Church of the Brethren	Alexander Mack (ca. 1708)
Moravians	Count Zinzendorf (ca. 1727)
Calvinistic Methodist	Howell Harris (ca. 1735)
American Dutch Reformed	Theodore Frelinghuysen (ca. 1737)
Seceders	Ebenezer Erskine (ca. 1740)
Shakers	Ann Lee (ca. 1741)
Methodists	John Wesley (ca. 1744)
Universalists	John Murray (ca. 1779)
Episcopalians	Samuel Seabury (ca. 1784)
African Methodist Episcopal Zion Church	Richard Allen (ca. 1787)
Harmony Society Church	George Rapp (ca. 1803)
Mormons	Joseph Smith (ca. 1829)
Disciples of Christ	Barton W. Stone / Alexander Campbell (ca. 1832)
Seventh Day Adventist	William Miller (ca. 1844) / Ellen G. White
Christadelphians	John Thomas (ca. 1848)
Christian Reformed	Gysbert Haan (ca. 1857)
Salvation Army	William Booth (ca. 1865)
Christian Scientists	Mary Baker Eddy (ca. 1879)
Jehovah's Witnesses	Charles Taze Russell (ca. 1884)
Nazarenes	Phineas Bresee (ca. 1895)

Denomination	Founder
Pentecostals	C.F. Parham / William Seymour / A.J.Tomlinson (ca.1903/1906)
Alliance	Albert Benjamin Simpson (ca. 1905)
Church of God in Christ	Charles Mason (ca. 1907)
Foursquare	Aimee Semple McPherson (ca. 1918)
Church of God	Joseph Marsh (ca. 1920)
Worldwide Church of God	Herbert W. Armstrong (ca. 1934)
Confessing Church	Martin Niemoller (ca. 1934)
Evangelical Free	E. A. Halleen (ca. 1950)
Moonies	Sun Myung Moon (ca. 1954)
Children of God	David Mo Berg (ca. 1969)
Universal Church of the Kingdom of God	Macedo de Bezarra (1977)

Obviously we cannot name all 33,000 Protestant denominations and their founders, nor the 150,000 plus pseudo-Christian denominations. But I think the point is obvious.

Only one Church was founded upon Christ in the year 33 AD, the Catholic Church. As Ignatius of Antioch, the disciple of John and friend of Peter and Paul, and the one referred to in Mark 9:35, declares in his letter to the *Smyrneans* (8): "[W]herever Jesus Christ is, there is the Catholic Church."

Irenaeus, the disciple of Polycarp, who in turn was the disciple of the apostle John, also makes mention of the Catholic Church as the authentic Church founded by Christ:

The blessed apostles [Peter and Paul] having founded and built up the Church of [Rome], handed over the office of the episcopate to Linus. Paul makes mention of this Linus in the Epistle to Timothy. To him succeeded Anecletus; and after him, in the third place from the apostles, Clement was chosen for the episcopate. He had seen the blessed apostles and was acquainted with them, and had their traditions before his eyes (Against Heresies, 3:3, trans. Jurgens).

History is on the side of the Catholic Church. No one can question with any sense of respectability the founding of the Catholic Church by Christ. Look at any encyclopedia in the world and you will find as the founder of the Catholic Church, Jesus Christ.

The Church was founded by Christ and his apostles, and it continues today through, with, and in Christ and the successors of the apostles, the bishops (cf. Eph. 2:20).

- Offshoots of the Lutherans include the Lutheran Brethren, the Evangelical Covenant Church, the Evangelical Free Church, the Evangelical Lutheran Church in America, the Missouri Synod Lutherans, the Wisconsin Synod Lutherans, and the Moravian Church.

- Offshoots of the Anabaptists include the North American Baptist Church, the Advent Christian Church, the Seventh Day Adventist Church, the Amish, the Conservative Mennonites, the General Conference of Mennonites, the Old Mennonite Church, the Brethren in Christ, the Hutterite Brethren, the Independent Brethren, and the Mennonite Brethren.

- Offshoots of the Anglican Church include the United Church of Christ, the Free Will Baptists, the Conservative Baptists, the Progressive National Baptists, the American Baptists, the Independent Bible Churches, the Friends United, the Friends General Conference, the United Methodists, the African Methodists, the Episcopal, the Free Methodists, and the many offshoots of the Pentecostal churches.

- Offshoots of Calvinism include the Presbyterian Church in America, the Presbyterian Church in the USA, the Orthodox Presbyterian Church, the Reformed Presbyterian Church, the Reformed Church in America, the Christian Reformed, the Churches of Christ, the Disciples of Christ, and the "Christian Churches."

The Catholic Church has many rites, yet one faith that traces itself back to Jesus Christ, the apostles and their successors, the bishops. Their Catholic identity is found in their union to the successor of St. Peter, the pope, in proclaiming the one true faith—in diverse cultural expressions—of Jesus Christ. Whether one is a member of the Roman, the Mozarabic, the Ambrosian, the Byzantine, the Chaldean, the Syro-Malabarese, the Alexandrian, the Coptic, the Abyssinian, the Antiochene, the Malankarese, the Maronite, or the Armenian rite, one is a member of the one Catholic Church founded by Jesus Christ through his apostles.

Was Constantine the founder of the Catholic Church?

Some have tried to make the emperor Constantine the founder of the Catholic Church. No historian from any reputable university accepts this.

Constantine was instrumental in calling together the bishops to meet at the Council of Nicaea (ca. 325). Yet sadly to say, he died denying the very teachings of the council he helped to call together. Instead of receiving baptism from a Catholic bishop, he was baptized by a heretical Arian bishop, Eusebius of Nicomedia. Constantine's death gave rebirth to and strength to

Arian Christianity. Constantine's successors would bring Arian Christianity into an even stronger conflict with Catholic Christianity.

It is quite clear that Constantine could never be considered the founder of the Catholic Church, but even if someone were to concede that Constantine was the founder of the Catholic Church (ca. 325) then you would have another problem to deal with. You would have to say that the Bible that all Christians cherish is the result of a Constantinian Church, since the Bible does not get put together until the late fourth century. Instead of a Christian Bible you would have to say we cherish a Constantinian Bible. This is obviously absurd.

The Bible was put together at the Councils of Hippo and Carthage III and IV by the pope and the bishops—the first list of books being initially approved by Pope Damasus in 382 and reaffirmed by Pope Innocent I in 411.

History and historians acknowledge that the Catholic (universal) Church was founded by Christ—the name "catholic" being officially written on paper for the first time in 107 AD by Ignatius of Antioch, a disciple of the apostle John and bishop of Antioch through ordination by the apostles Peter and Paul.

Cyril of Jerusalem (ca. 350) in his *Catechetical Instructions* reminds us why the Catholic Church is the only true Church:

The Church is called Catholic or universal because it has spread throughout the entire world, from one end of the earth to the other. Again, it is called Catholic because it teaches fully and unfailingly all the doctrines which ought to be brought to men's knowledge, whether they are concerned with visible or invisible things, with the realities of heaven or the things of the earth. Another reason for the name Catholic is that the Church brings under religious obedience all classes of men, rulers and subjects, learned and unlettered. Finally, it deserves the title Catholic because it heals and cures unrestrictedly every type of sin that can be committed in soul or in body, and because it possesses within itself every kind of virtue that can be named, whether exercised in actions or in words or in some kind of spiritual charism. (Cf. Cat. 18: 23-25: PG 33, 1043-1047).

Is the Catholic Church the "Whore of Babylon"?

Some fringe or radical Protestants believe that the book of Revelation refers to the Catholic Church in Rome as the "Whore of Babylon" with her seven heads. They often like to cite Revelation 17:1-18.

The seven heads are a reference to the legendary Seven Hills of Rome upon which the "whore" is found. This cannot be a reference to the Catholic Church or the pope, since no pope has ever been seated on any of the Seven Hills of Rome. No pope has ever lived on the Capitoline, Palatine, Esquiline,

Aventine, or on the three little hills in central Rome which make up the seven hills. The seven hills and in particular the three little hills are where the pagan religions and the Roman governments were situated. When John wrote the book of Revelation, the popes and the Catholics lived in Trastevere, a district across the Tiber River and away from the city. Hence, the Catholic Church could never be associated with the "Whore of Babylon." And even today, the Lateran (the pope's Church) and the Vatican (where the pope lives) could never be associated with the traditional Seven Hills of Rome.

The sacred author of Revelation was referring to pagan Rome under the leadership of the emperor Nero who persecuted Christians, killing the apostles Peter and Paul. Or for some scholars, it is a reference to the emperor Domitian who brutally persecuted Catholics in imitation of Nero. Some referred to Domitian as the "re-incarnation" of Nero.

Nero and Domitian persecuted the Church from these hills. The Capitoline was the religious and political center of the empire and the Palatine was where the imperial palace was situated.

The historical Babylon persecuted the Jews, the People of God, between 610 and 538 BC. The Babylonians destroyed the temple and sent the Jews into exile. The Romans became the modern day version of the Babylonians by destroying Jerusalem and the temple in 70 AD. To this very day, the temple has not been reconstructed.

Jesus inaugurated the establishment of the new people of God—made up of Jews and Gentiles—in a Church. The Romans were now persecuting the Church, the People of God. Revelation used symbolic language, or apocalyptic language, in order to encourage Christians to persevere through their struggle with the Roman authorities (see also 1 Pet. 5:13). Jews and Christians were quite aware that Babylon was a hidden reference to Rome.

Finally if there is any doubt as to the distinction between the Catholic Church in Rome and the Roman Empire, all you have to do is look to the words of Ignatius of Antioch, the disciple of the apostle John and the friend of Peter and Paul. This is what he thought of the "Catholic" Church in Rome:

Ignatius, who is also called Theophorus, to the Church which has found mercy, through the majesty of the Most High Father, and Jesus Christ, His only-begotten Son; the Church that is beloved and enlightened... the Church that presides in the capital of the Romans, worthy of God, worthy of honor, worthy of highest happiness, worthy of praise, worthy of obtaining her every desire, worthy of being deemed holy, the Church that presides in love, named from Christ and from the Father.... (Address to the Romans).

Does this sound like the Catholic Church is the "Whore of Babylon"? Does St. Paul's *Letter to the Romans* in the Bible sound like the Church in Rome is a "whore?"

Was there a great apostasy?

Mormons claim that they are part of a "restored Church." They claim that the early Church fell away from the truth of Jesus Christ. They often like to quote Acts 20:29-30, 2 Thessalonians 2:1-3, 2 Peter 2:1 and Matthew 7:15. The only problem with these quotes is that they in no way refer to a total apostasy. In fact, there are no quotes in the Scriptures that point to a complete apostasy in the Church. All the examples of apostasy are examples of individual members or groups committing the sin of apostasy.

To accept the theory of the "great apostasy" is to make Jesus a liar, for he said, "On this rock I will build my Church and the gates of hell shall not prevail against it" (Mt. 16:18). He also reminded us: "Behold, I am with you always, until the end of the age" (Mt. 28:20). To believe in a great apostasy is to believe that Christ would have abandoned us and allowed the gates of hell to prevail against the Church! This cannot be. The Bible tells me so!

It is the Church that is the "Church of the living God, the pillar and foundation of truth" (1 Tim. 3:15). It is the Church that is the pillar and foundation of truth, not the Bible, for the Bible flowed from the Church and is interpreted by the Church. Jesus built his Church on a strong foundation that would never fall (cf. Mt. 7:24-27). And he entrusted his Church with the power of running it according to his will (Mt. 18:15-18). If the Church is the foundation of truth built on a foundation that will never fail, a foundation upon which the gates of hell will not prevail against it, then how in the world could the Church remain true to its mission if there was a great apostasy! It is absurd to think that only with the life of the founder of Mormonism, Joseph Smith, in 1829, that this "great apostasy" would end. It is absurd to think that God would allow his Church to live in apostasy for 19 centuries.

The history of Christianity never, ever makes mention of a so-called "great apostasy." No early Christian writer, non-Christian writer, or hostile opponent of Christianity ever makes mentions of such an apostasy. If an apostasy took place why are there no enemies of the Church pointing to this so-called apostasy in history. And if there was an apostasy, who began it? When did it begin and with whom?

Finally, the same arguments that point to the impossibility of Constantine being the founder of the Catholic Church also apply to those who claim that the early Church experienced a great apostasy. If a great apostasy occurred, then Mormons would be worshiping out of the "Great Apostasy Bible."

- The Bible was put together at the Councils of Hippo and Carthage III and IV by the pope and the bishops—the first list of books being initially approved by Pope Damasus in 382 and reaffirmed by Pope Innocent I in 411.

Who founded the Church in Rome?

In a recent television program the "Roman Church" was attacked as being un-Christian. Anything Roman or associated with Rome, from the point of view of these individuals, is not Christian.

Besides the fact that the apostle Paul was a Roman citizen, history shows us quite the opposite. Tertullian in 193 wrote:

> *In this chair in which he himself sat, Peter,*
> *In mighty Rome, commanded Linus, the first elected, to sit down.*
> *After him, Cletus too accepted the flock of the fold.*
> *As his successor, Anacletus was elected by lot.*
> *Clement follows him, well-known to apostolic men.*
> *After him Evaristus ruled the flock without crime.*
> *Alexander, sixth in succession, commends the fold to Sixtus.*
> *After his illustrious times were completed, he passed it on to Telesphorus.*
> *He was excellent, a faithful martyr.*
> *After him, learned in the law and a sure teacher,*
> *Hyginus, in the ninth place, now accepted the chair.*
> *Then Pius, after him, whose blood-brother was Hermas,*
> *An angelic shepherd, because he spoke the words delivered to him;*
> *And Anicetus accepted his lot in pious succession.*

> *Adversus Marcionem libri quinque* , 3, 276-285; 293-296

Some groups like to attack the importance of the Church of Rome by making the claim that Peter could not have been the first pope since there is no evidence of his presence in Rome. This is an absurdity to all historians. The very bones of Peter are found under the altar of St. Peter's Basilica in Rome. It has always been accepted that the apostles Peter and Paul founded the Church in Rome. Virtually all Protestants recognize this reality as well. It is only fringe groups that like to deny this.

The first great historian of Christianity, Eusebius (ca. 324), writes in his *Ecclesiastical History* (3:3, 3:25) the following:

> *It is related that in his time Paul was beheaded in Rome itself, and that Peter likewise was crucified, and the title "Peter and Paul," which is still given to the cemeteries there confirms the story, no less than does a writer of the Church named Caius.... Caius...speaks...of the places where the sacred relics of the apostles in question are deposited: 'But I can point out the trophies of the apostles, for if you will go to the Vatican or to the Ostian Way you will find the trophies of those who founded this Church.'*

It is clear that the Church of Rome was founded by Peter and Paul, with Peter being the first pope. On Vatican Hill, the very place of Peter's death was built St. Peter's Basilica and Vatican City.

- Peter was crucified where the front doors of St. Peter's Basilica now stand. Before the building of the Basilica, an ancient Roman Coliseum for gladiatorial games was present on that site.

Peter, the Rock upon which Jesus built his Church!

[Jesus said,] "Who do men say that the Son of man is?"... Simon Peter replied, "You are the Christ, the Son of the living God." And Jesus answered him, "Blessed are you, Simon Bar-Jona! For flesh and blood has not revealed this to you, but my Father who is in heaven. And I tell you, you are Peter, and on this rock I will build my church, and the powers of death [of hell] shall not prevail against it. I will give you the keys of the kingdom of heaven and whatever you bind on earth shall be bound in heaven, and whatever you loose on earth shall be loosed in heaven." (Mt. 16: 13-19, RSV).

There is only one Church that can trace itself back to Peter and thus to Jesus—the Catholic Church. As we saw previously, Peter was the first pope, followed by Linus, Anecletus, Clement, and so forth. This line of succession goes on all the way to our current pope—one of the greatest unbroken lines of succession in world history. This alone is a miracle.

The gift of the "keys of the kingdom" to Peter by Jesus is quite significant. It finds its origins in Jewish history and is cited in Isaiah 22:21-22. The keys were given to the chief official within the Kingdom of David. It was a symbol of authority on behalf of the king and was an office that did not end with the death of the official. When the office became vacated another successor took his place. Jesus intended for Peter to have successors, and in fact history proves it!

It is also worth noting that at the time of the writing of Matthew's Gospel (ca. 70 AD) Peter had already been killed by being crucified upside down on Vatican Hill (ca. 67AD). The claim by Jesus that he would build his Church upon Peter, consequently, implies that there would be successors to Peter with the authority of Peter; otherwise, it would have been somewhat odd to place such a text within a Gospel when the person being written about is dead.

It is also interesting to note that the Gospel of Mark is seen by scholars as being at the heart or core of all the other Gospels. This is interesting since Peter dictated this Gospel to Mark!

Another note worth mentioning is that the name Peter, "Rock," had never been used before as a name for someone. It is an odd name to give to anyone. Christ had a purpose for this odd name!

Now let us look at the arguments that Protestants use against this text in order to try to undermine its importance, for if we accept this text, then we have no choice but to recognize the true Church as that founded upon Peter. We have no other option than to recognize the true Church as the Catholic Church.

Many Protestants like to point out the distinction between the Greek *Petros* and *Petra*. They point out that in the Greek text of the Gospel of Matthew, Jesus changes Simon's name to *Petros*, which can at times mean "chip," even though the two Greek words—*Petros and Petra*—are often used interchangeably for the word "rock." They claim that Jesus was referring to Peter as a chip off the block. If he had wanted to name him the "Rock" he would have used *Petra*, which has no other interpretation than "Rock."

This argument at first may appear appealing, but it is terribly flawed. First, the reason why Matthew used the Greek word *Petros* as opposed to *Petra* to name Simon Peter is that *Petros* is a masculine proper noun, whereas *Petra* is a feminine proper noun. Matthew would not have Jesus calling Simon Peter a feminine "Rock." Hence, Matthew uses *Petros*, the masculine proper noun.

But all of this is of little significance, since Jesus never spoke Greek. Jesus spoke Aramaic, and so Jesus would not have used either word. It is also worth mentioning that the original text of Matthew, albeit lost to history, was in Aramaic. In any case, Jesus would have used the word *Kepa* which means one and only one thing, "Rock." Therefore, Jesus' exact words to Simon Peter would have been, "You are *Kepa*, and on this *Kepa* I will build my Church." In other words, "Your name is Rock and on this Rock I will build my Church and the gates of hell shall not prevail against it." After 2000 years, the gates of hell have not prevailed against his glorious Church, despite many attacks.

Christ built his Church on *Kepa*. Paul's writings in the New Testament always refer to Peter in the Greek transliteration of the Aramaic *Kepa*, *Kephas*. Paul, with one exception (Gal. 2:7-8), always refers to Peter as *Kephas*, which means one and only one thing, "Rock."

Saint Leo the Great (ca. 461) succinctly summarizes the reality of Peter as "the Rock" when he paraphrases and develops the implications of the words of Jesus:

You are Peter: though I [God] am the inviolable rock, the cornerstone..., the foundation apart from which no one can lay any other, yet you also are a rock, for you are given solidity by my strength, so that which is my very own because of my power is common between us...by participation (Cf, Sermon 4 *de natali ipsius,* 2-3: PL 54,149-151).

Are the popes antichrists?

Some fundamentalist denominations like to argue that the papacy is the seat of the antichrist. They take the number of the beast, 666, from Revelation 13:18, and use the Latin letters, which also represent numbers, to come up with the number 666 for the pope; that is, by counting up the letters that make up the phrase *Vicar of the Son of God, Vicarius Filii Dei*, they come up with 666.

The only problem with such an assertion is that the pope is never called, nor has he ever been called, by the title *Vicar of the Son of God*. His appropriate title is *Vicar of Christ, Vicarius Christi*, which in no way adds up to 666.

If you play games such as this, you can almost make anyone turn out to be the antichrist. All you need to do is to make up a title that fits the designation. This was a common practice during the Protestant Reformation; almost everyone, including Luther, was accused of bearing the number of the beast.

Another flaw with this argument is that the book of Revelation was not written in Latin. It was written in Greek, and it was not Latin numerology that was being used, but Hebrew numerology. For example, the Greek form of the name Nero Caesar in Hebrew numerology is *nrwn qsr*, which adds up to 666 (n=50; r=200; w=6; q=100; s=60) (50+200+6+50+100+60+200=666). When we use the Latin form of the name Nero Caesar in Hebrew letters, *nrw qsr*, we end up with 616. This is extremely interesting, since as was mentioned earlier, no two ancient manuscripts of the New Testament are exactly alike. Given this reality, when we look at differing manuscripts, we notice that some of the ancient manuscripts give the number for the antichrist as 666 and others 616—both designating Nero Caesar as the antichrist. This would make a lot of sense since Nero Caesar was known to have begun the first great persecution against the Catholics which resulted in the deaths of Peter and Paul in Rome.

The "whore of Babylon" (Rev. 17:1-6, 9) symbolizes pagan Rome and the antichrist is the persecuting Nero.

Some scholars place the persecution of the Christians, referred to in the book of Revelation, during the reign of Domitian (81-96) when atrocious persecutions took place. Nero's name would have been used to claim that Domitian was another Nero, another antichrist.

In any event, Revelation was a book written for the encouragement of the faithful in times of persecution. Its relevance for us today is that we as Christians are also persecuted, albeit in mostly subtle ways, and we too must fight and persevere against the antichrists of today. Just as in Perganum, where a throne to Satan was erected (Rev. 2:13), many of today's cities and nations have decided to raise thrones to the ways of Satan.

Why is the pope so important?

Had Alexandria triumphed and not Rome, the extravagant and muddled stories [of the Gnostics] would be...perfectly ordinary.
Jorge Luis Borges

Popes have always exercised supreme authority in honor and jurisdiction in Christianity. From Peter to the current pope, the Church has always recognized this reality.

Peter (33-67) arranged for the successor of Judas, Matthias (Acts 1:25f), presided over the first council of the Church in Jerusalem, and admitted Gentiles into the Church (Acts 15). Linus (67-76) developed the clergy in Rome. Anacletus (76-88) was consulted regarding the proper consecration of bishops, and Clement (88-97) was called upon to squash the disobedience of the Corinthians. Alexander I (105-115) issued the decree that unleavened bread was to be used for consecration; Sixtus I (115-125) decreed the praying of the *Sanctus* and Telesphorus (125-136) the praying of the *Gloria*. Pius I (140-155) issued the decree regarding the proper date for the celebration of Easter. Hyginus (136-140) was asked to squash the heresy of Gnosticism, Anicetus (155-166) the heresy of Manichaeism, Soter (166-175) the heresy of Montanism, and Victor I (189-199) the heresy of Adoptionism. Damasus I (366-384) chose which books would be in the Bible and which would not. The popes have led the way for 2000 years.

Whenever the Church sought guidance, it always looked to the successor of Peter, the pope. This pattern continued and continues uninterrupted to this very day! Vatican I would officially affirm this pattern under the doctrine of Papal Infallibility.

At the Council of Chalcedon (451) Pope Leo's letter regarding Christ's two natures was read. After Leo's affirmation that Christ was fully human, fully divine, without any confusion, change or division amongst his natures, the bishops sprang to their feet and proclaimed: "Peter has spoken through Leo." The successor of Peter had led the bishops and all the Christian faithful to the truth.

The Holy Father, the pope, is important because he is the successor of the apostle Peter who was entrusted with the keys to the Kingdom and who was entrusted to lead the Church (Mt. 16:18f). In rabbinic terminology the ability to "bind and loose" (cf. Mt. 16:18f) is equated with the authority to decide what is allowed or forbidden by law as well as the authority to include and exclude individuals from a community.

In the naming of the apostles, Peter is always named at the head of the list (Mt. 10:1-4; Mk. 3:16-19; Lk. 6:14-16; Acts 1:13). Of all the apostles Peter is named 195 times in the New Testament, whereas the next most often mentioned apostle, John, is only mentioned 29 times. Peter is also the one who usually spoke as the representative of the apostles (Mt. 18:21; Mk. 8:29;

Lk. 9:32; 12:41; Jn. 6:69). (The Bible emphasizes Peter's authority and special role by phrases such as "Simon Peter and the rest of the apostles" or simply "Peter and his companions" (Lk. 9:32; Mk. 16:7; Acts 2:37).) It is to Peter that an angel is sent to announce the resurrection of Jesus (Mk. 16:7). It is to Peter that the risen Christ appears to before appearing to the other apostles (Lk. 24:33-35). It is Peter that leads the apostles in selecting the replacement for Judas with Matthias (Acts 1:15-26). It is Peter who was called upon to strengthen his brothers in the faith (Lk. 22:31-32). Peter is the one who preached to the crowds at Pentecost as the leader of the apostles (Acts 2:14-40) and received the first converts (Acts 2:41). It is Peter who performed the first miracle after the resurrection (Acts 3:6-7) and it is Peter who inflicted the first punishment on the disobedient, on Ananias and Saphira (Acts 5:1-11). It is Peter who excommunicated the first heretic, Simon Magnus (Acts 8:21). It is Peter who led the Church's first council, the Council of Jerusalem, and encouraged the baptism of the Gentiles (Acts 10:46-48). It is Peter who pronounced from the council the first dogmatic decision (Acts 15:17). And it is to Peter that Paul went to make sure his teachings were in line with his (Gal. 1:18). And finally, it is Peter alone who was told before Jesus' Ascension into heaven to nourish the faithful in the faith (Jn. 21:15-17), even though the other apostles were present in their midst.

Because of these realities, the pope is in charge of leading the Church. All are to be obedient to him in faith and morals and in respect. He is infallible in and by himself in the areas of faith and morals when he speaks *ex cathedra*; that is, when he speaks for the universal Church, from the authority of Peter, with the clear indication that what he is to say is to be held infallible. He also speaks infallibly in an "ordinary" manner when he affirms a teaching that has always been held by the Church (i.e., Pope John Paul II reaffirmed two thousand years of Christianity when he stated that the ordination of women is not possible since it is not within the "deposit of the faith.").

Bishops share in this infallibility when they teach in union with the pope.

Let us look at the words of those who walked and talked and learned from the apostles in regards to papal authority.

The fourth pope, Clement of Rome, a convert of the apostle Peter, a friend of Peter and Paul, and the Clement mentioned in Philippians 4:3, makes the following statement (ca. 88-97):

> *The Church of God which sojourns in Rome to the Church of God which sojourns in Corinth.... (Letter to the Corinthians, Address, trans. Lake, vol. 1). Owing to the sudden and repeated misfortunes and calamities which have befallen us, we consider that our attention has been somewhat delayed in turning to the questions disputed among you.... (Ibid., 1).*

Clearly the fourth pope is asserting his authority as the successor of Peter. He is addressing, in his letter to the Corinthian community, the abuses to the Gospel that are taking place there.

Why would the bishop of Rome be interfering in the affairs of the Corinthian community? Clearly, he was doing so by authority. The Corinthian community, upon reading the letter, and acknowledging its supreme authority, made it part of their liturgical readings. In fact, this letter would almost take on canonical status, and in fact in some communities it did take on canonical status. It was not until the decisions of the fourth century popes and bishops that this letter would be deemed as not having canonical status. It may not have been Scripture, but it certainly was understood as authoritative in faith and morals.

Let us look at Ignatius of Antioch (ca. 107), the disciple of John, and his understanding of papal authority:

> *Ignatius, who is also called Theophorus, to the Church which has found mercy, through the majesty of the Most High Father, and Jesus Christ, His only-begotten Son; the Church that is beloved and enlightened...the Church that presides in the capital of the Romans, worthy of God, worthy of honor, worthy of the highest happiness, worthy of praise, worthy of obtaining her every desire, worthy of being deemed holy, the Church that presides in love, named from Christ and from the Father....* (*Letter to the Romans*, trans. Lake).

Irenaeus of Lyon, a disciple of Polycarp, who in turn was a disciple of John, writes in the year 202 AD: "All other churches must bring themselves into line with the Church that resides in Rome, on account of its superior authority" (CCC 834).

If this is not the recognition of papal primacy, what is? The ancient writer Hermas, the man mentioned in Romans 16:14, in "visions" (2) recognized the importance of Rome when he made sure that some key writings were sent to Pope Clement for approval. As he writes:

> *[You] shall write two little books and send one to Clement [the successor of the apostle Peter].... Clement shall then send it to the cities abroad, because that is his duty....*

And in Cyprian of Carthage (ca. 251) we read in *De Ecclesiae Unitate* (cf. 2-7):

> *The blessed apostle Paul teaches us that the Church is one, for it has 'one body, one spirit, one hope, one faith, one baptism, and one God.' Furthermore, it is on Peter that Jesus built his Church, and to*

him he gives the command to feed the sheep; and although he assigns like power to all the apostles, yet he founded a single chair, and he established by his own authority a source and an intrinsic reason for that unity. Indeed, the others were that also which Peter was; but a primacy is given to Peter, whereby it is made clear that there is but one Church and one Chair—the Chair of Peter. So too, all are shepherds, and the flock is shown to be one, fed by all the apostles in single-minded accord. If someone does not hold fast to this unity of Peter, can he imagine that he still holds the faith? If he deserts the chair of Peter upon whom the Church was built, can he still be confident that he is in the Church?

In his *Letter to all his People* [43 (40) 5] written in 251 AD Cyprian reminded his people that the faith of the pope is the faith of the Church:

They who have not peace themselves now offer peace to others. They who have withdrawn from the Church promise to lead back and to recall the lapsed to the Church. There is one God and one Christ, and one Church, and one Chair founded on Peter by the word of the Lord. It is not possible to set up another altar or for there to be another priesthood besides that one altar and that one priesthood. Whoever has gathered elsewhere is scattering.

Obviously, from what we see in these earliest of writings, papal authority was well recognized throughout Christianity.

The first seven centuries of Christianity was divided between Catholic Christianity and Arian Christianity (which denied the Trinity). Catholic Christianity was always supported by the popes, and Arian Christianity almost always found support from the emperors of the East and the bishops seeking the favor of these emperors.

Catholic Christianity would prevail under the influence of the popes. Had it not, Christianity would be radically different than it is today!

- One of the great ironies of history is that Martin Luther, the first Protestant, was a priest in the Augustinian order before his break. His spiritual father in faith, St. Augustine of Hippo, after Rome had spoken regarding the Pelagian heresy, concluded (*Sermons* 131:10), "The matter is at an end." If only Luther would have recognized Rome's authority as much as the founder of his order did, Christianity would be quite different.

Without the popes, the successors of St. Peter, there would be no authentic Christianity!

In 1517, the founder of Protestantism, Martin Luther, had a meeting with the person who would eventually become known as the "father" of English and American Protestantism, the lawyer John Calvin. They met to iron out their theological differences. They could come to no agreement. At one point, in frustration, Luther turned to Calvin and said, "I started all this and you should follow what I started!" Calvin retorted, "Who in the world do you think you are, the pope?" In this ironic retort was a truth that these protestors had not realized—that without an ultimate *decision maker* there could be no consensus on religious belief. President Truman used to say, "The buck stops here!" In other words, the ultimate and final decisions stop with him. So too, the ultimate and final decisions stop with the successor of Peter, the pope. Otherwise, Christianity would simply be an accumulation of confusing beliefs and practices. Truth would be left to personal opinion! And that is a religion doomed to die!

Let us examine the popes of the first five centuries, in particular those popes that taught doctrines that all mainline Christians take for granted. When we do this we can see that without the popes, Christianity would not exist or would exist as simply an accumulation of confusing beliefs!

The first pope, Peter led the first council of the Church, the Council of Jerusalem, and admitted the Gentiles into the Church. Peter chose his successor: *"After the Holy Apostles Peter and Paul had founded and set up the Church in Rome they gave over the exercise of the office of bishop of Rome to Linus"* (Irenaeus, *Adv. Haereses*, III, iii, 3). The first successor of Peter, the second pope, Linus (67-76), is chosen to lead the Church. He is best known for encouraging the growth of the clergy and the development of parishes to fulfill the spiritual needs of the growing Christian population. The ninth pope, Hyginus (136-140) instituted the use of godparents for infant baptisms. The tenth pope, Pius I (140-155), opposed agnosticism and established the date for Easter as the first Sunday after the March full moon. The eleventh pope, Anicetus (155-166), emphasized the celebration of Easter as the central Christian feast. The twelfth pope, Soter (166-175), affirmed the Sacrament of Matrimony. The twenty-first pope, Cornelius (251-253), opposed the heresy of Novatianism which believed that sins could not be forgiven and that the Church was solely made up of saints. The twenty-second pope, Lucius I (253-254), reiterated the ban on premarital sexual relationships and the living together before marriage. The twenty-sixth pope, Felix I (269-274), affirmed Christ as being God and man, as having two natures in one Person. The thirty-fifth pope, Julius I (337-352), decreed that Christmas should be celebrated on December 25. The thirty-sixth pope, Liberius (352-366), fought against the heresy of Arianism. The thirty seventh pope, Damasus I (366-384), decided which books would make up the Bible and which would not. He then had the Scriptures translated into the vernacular by Jerome. The books he included into the Bible are what all Christians use as their Bible of worship. The books he excluded include the following: The Gospel of Thomas, the Dialogue of the Savior, the Gospel of Mary, the Infancy Gospel of Thomas, the Infancy Gospel of James, the Gospel of Peter, the Gospel of Bartholomew, the Gospel of Nicodemus, the Gospel of the Nazoreans, the Gospel of the Ebionites, the Gospel of Philip, the Gospel of the Egyptians, and on and on the list goes. He excluded the Apocryphon of James, the Apocryphon of John, the Apocalypse of Paul, the two Apocalypses of James, the Apocalypse of Peter, the Acts of Peter and the Twelve Apostles, the Acts of Andrew, the Acts of John, the Acts of Thomas, etc., etc..

Is it not interesting that no denomination questions the work of Pope Damasus I. In a sense, they are accepting—albeit reluctantly, subliminally, or subconsciously—that when it came to putting the Bible together the pope was infallible.

What other teachings did the popes promote as the norm for Christianity? Remember these beliefs that the popes rejected or accepted were firmly rooted as "competing" versions of Christianity. Yet it is the version of Christianity defined and fought for by the popes that would win the day. It is that version (on the basic doctrines of Christianity) that mainline Christians would accept—even to this day! Why? As Catholics we would say, "Where the pope is there is the faith!"

It is the popes that condemned Docetism and Gnosticism, the beliefs that denied Jesus' humanity. It is the popes that condemned Marcionism, the belief that the Old Testament should be eliminated from the Scriptures. It is the popes that rejected Montanism, Donatism, and Novatianism, the beliefs that held that serious sins could never be forgiven. It is the popes that condemned Modalism, the belief that the Father, Son, and Holy Spirit were modes of one Divine Person—as opposed to three persons in one God. It is the popes who condemned Monarchianism which argued Jesus became divine after his baptism. It is the popes who rejected Subordinationism that viewed the Father and Son as unequal. It is the popes who rejected Sabellianism which denied the distinction between the Father and the Son. It is the popes that rejected Patripassionism which argued that it was the Father that was crucified and not the Son. It is the popes that condemned Manicheaism which held the belief in two competing and equal principles as rulers of the universe, one good, one bad, one matter, one spirit. It is the popes that rejected Arianism which denied Jesus' divinity and the Trinity. It is the popes which condemned Pneumatomachism which denied the divinity of the Holy Spirit. It is the popes that condemned Eunomianism which rejected the divinity of Christ and the Holy Spirit. It is the popes who rejected Priscillianism which denied the preexistence of the Son and denied the humanity of Jesus. It is the popes who condemned Monophysitism that claimed that Christ had only one nature. It is the popes that rejected Nestorianism which argued that Jesus was two distinct persons. It is the popes that rejected Pelagianism which denied "original sin" and argued that one could work oneself into heaven without grace. It is the popes that condemned Traducianism which argued that the soul was not created by God but by human beings. It is the popes who rejected Monothelitism which denied that Christ had a human and a divine will. It is the popes that rejected Albigensianism which viewed matter as evil and suicide as a way of freeing oneself from matter, from the body. It is the popes that rejected Cartharianism which renounced baptism and marriage and the holiness of the body. It is the popes which condemned Jansenism which argued that Christ did not die for all. The list goes on and on!

All mainline Christians, today, accept what these Catholic popes taught and fought for. Why do they accept these teachings and yet do not accept all of the popes' teachings? Why the picking and choosing? All the heretics above had the same material to draw upon? Yet they disagreed over what the Scriptures meant and what the Holy Spirit was saying! Why do mainline Christians accept the teachings of the popes if these above groups did not?

Why is apostolic succession so important?

The agreement among [Catholics] is astonishing and quite amazing...
Celsus (ca. 170) a renowned anti-Christian philosopher

The people of God always knew of the importance of spirit-filled successors to their leaders. When Moses' earthly journey was approaching its end, Moses went to Joshua and called the spirit of God upon him by the "laying on of hands" (cf. Ex. 34:1-12). Thus Joshua succeeded Moses in leading the people of God.

In a similar yet more tragic manner, after the death of Judas, the apostles sought out a successor to replace Judas. Two men were proposed to succeed Judas, Barsabbas and Matthias.

The apostles prayed and then cast lots. The lot fell on Matthias. Matthias then became the successor of Judas and took his place alongside the eleven (Acts 1:15-26).

Paul mentions how he "laid the foundation" for others, successors, to build upon (1 Cor. 3:10). Paul mentions Silvanus and Timothy as being ordained to the office of apostle and thus having apostolic authority (cf. 1 Thess. 1:1; 2:6,7; 2 Tim. 1:6). Other examples of passing on the teaching authority of the apostles through apostolic succession and the "laying on of hands" can be seen and implied in Acts 14:23, Acts 20:28, 1 Corinthians 12:27-29, Ephesians 2:20; 4:11, and 1 Timothy 3:1-8; 4:13-14; 5:17-22.

Apostolic succession is that reality that allows the Church of the year 2007 AD to be connected to the faith of the Church of 33 AD. Each Catholic bishop in the world can trace his authority from bishop to bishop all the way down to the apostles themselves.

The apostles did not live in a vacuum. They walked and talked with people and in turn they appointed men to take their place as bishops in guiding their communities, and these bishops were in turn succeeded by other bishops in the same line of succession (cf. Acts 1:15-26; i.e., Matthias succeeded Judas in the office of apostle).

In this way the deposit of the faith would always be protected. The faith of Christ would always be kept pure. Without apostolic succession, Christianity would be a mist of confusion. It is for this reason that Paul instructs Timothy to choose successors with caution (1 Tim. 5:22; 2 Tim. 2:2).

Let us always be faithful to the successors of the apostles, the bishops in union with the pope, the successor of the leader of the apostles.

Irenaeus (ca. 140), as we remember from above, was a pupil of Polycarp (ca. 69-156), and Polycarp was a disciple of the apostle John. This bears repeating because it is a witness to the authority of this man. Irenaeus describes the importance of noting the successors of the churches [i.e., dioceses, communities].

> *The blessed apostles Peter and Paul, having found and built up the Church of Rome, handed over the office of the episcopate to Linus. Paul makes mention of this Linus in the Epistle to Timothy [2 Tim. 4:21]. To him succeeded Anecletus; and after him, in the third place from the apostles, Clement was chosen for the episcopate. He had seen the blessed apostles and was acquainted with them, and had their traditions before his eyes* (Against Heresies, 3,3, trans. Jurgens, vol. 1).

Clement of Rome, Peter's friend and successor, makes it quite clear how important apostolic succession is:

> *Our apostles knew through our Lord Jesus Christ that there would be strife for the title of bishop. For this cause, therefore, since they had received perfect foreknowledge, they appointed those who [were properly chosen], and afterwards added the codicil that if they should fall asleep [that is, die], other approved men should succeed to their ministry* (Letter to the Corinthians, 44, trans. Lake).

Let us always be faithful to the successors of the apostles! And if there is any question about proving such a succession I place before you the successors of Peter that can be found in any history book or encyclopedia.

This line of succession is unequalled in the world. In fact it is miraculous. This very miracle is a proof of the importance of apostolic succession, the primacy of Peter, and the truth that the "gates of hell" would never prevail against the Church (Mt. 16:18f).

Peter (64 or 67)
Linus (67-76)
Anacletus (76-88)
Clement (88-97)
Evaristus (97-105)
Alexander I (105-115)
Sixtus I (115-125)
Telesphorus (125-136)
Hyginus (136-140)
Pius I (140-155)
Anicetus (155-166)
Soter (166-175)
Eleutherius (175-189)
Victor I (189-199)
Zephyrinus (199-217)
Calixtus I (217-222)
Urban I (222-230)
Pontian (230-235)
Anterus (235-236)
Fabian (236-250)
Cornelius (251-253)
Lucius I (253-254)
Stephen I (254-257)
Sixtus II (257-258)
Dionysius (259-268)
Felix I (269-274)
Eutychian (275-283)
Caius (283-296)
Marcellinus (296-304)
Marcellus I (308-309)
Eusebius (309-310)
Melchiades (311-314)
Sylvester I (314-335)
Mark (336)
Julius I (337-352)
Liberius (352-366)
Damasus I (366-384)
Siricius (384-399)
Anastasius I (399-401)

Innocent I (401-417)
Zosimus (417-418)
Boniface I (418-422)
Celestine I (422-432)
Sixtus III (432-440)
Leo I (440-461)
Hilary (461-468)
Simplicius (468-483)
Felix III (II) (483-492)
Gelasius I (492-496)
Anastasius II (496-498)
Symmachus (498-514)
Hormisdas (514-523)
John I (523-526)
Felix IV (III) (526-530)
Boniface II (530-532)
John II (533-535)
Agapitus I (535-536)
Silverius (536-537)
Vigilius (537-555)
Pelagius I (556-561)
John III (561-574)
Benedict I (574-579)
Pelagius II (579-590)
Gregory I (590-604)
Sabinian (604-606)
Boniface III (607)
Boniface IV (608-615)
Deusdedit (Adeodatus I) (615-618)
Boniface V (619-625)
Honorius I (625-638)
Severinus (640)
John IV (640-642)
Theodore I (642-649)
Martin I (649-655)
Eugenius I (654-657)
Vitalian (657-672)
Adeodatus II (672-676)
Donus (676-678)

Agatho (678-681)
Leo II (682-683)
Benedict II (684-685)
John V (685-686)
Conon (686-687)
Sergius I (687-701)
John VI (701-705)
John VII (705-707)
Sisinnius (708-708)
Constantine (708-715)
Gregory II (715-731)
Gregory III (731-741)
Zachary (741-752)
Stephen II (III) (752-757)
Paul I (757-767)
Stephen III (IV) (768-772)
Adrian I (772-795)
Leo III (795-816)
Stephen IV (V) (816-817)
Paschal I (817-824)
Eugenius II (824-827)
Valentine (827)
Gregory IV (827-844)
Sergius II (844-847)
Leo IV (847-855)
Benedict III (855-858)
Nicholas I (858-867)
Adrian II (867-872)
John VIII (872-882)
Marinus I (882-884)
Adrian III (884-885)
Stephen V (VI) (885-891)
Formosus (891-896)
Boniface VI (896)
Stephen VI (VII) (896-897)
Romanus (897)
Theodore II (897)
John IX (898-900)
Benedict IV (900-903)
Leo V (903)
Sergius III (904-911)
Anastasius III (911-913)
Landus (913-914)

John X (914-928)
Leo VI (928)
Stephen VII (VIII) (928-931)
John XI (931-935)
Leo VII (936-939)
Stephen VIII (IX) (939-942)
Marinus II (942-946)
Agapitus II (946-955)
John XII (956-964)
Leo VIII (963-965)
Benedict V (964-966)
John XIII (965-972)
Benedict VI (973-974)
Benedict VII (974-983)
John XIV (983-984)
John XV (985-996)
Gregory V (996-999)
Sylvester II (999-1003)
John XVII (1003)
John XVIII (1004-1009)
Sergius IV (1009-1012)
Benedict VIII (1012-1024)
John XIX (1024-1032)
Benedict IX (1032-1044)
Sylvester III (1045)
Benedict IX (1045)
Gregory VI (1045-1046)
Clement II (1046-1047)
Benedict IX (1047-1048)
Damasus II (1048)
Leo IX (1049-1054)
Victor II (1055-1057)
Stephen IX (X) (1057-1058)
Nicholas II (1059-1061)
Alexander II (1061-1073)
Gregory VII (1073-1085)
Victor III (1086-1087)
Urban II (1088-1099)
Paschal II (1099-1118)
Gelasius II (1118-1119)
Calixtus II (1119-1124)
Honorius II (1124-1130)
Innocent II (1130-1143)

Celestine II (1143-1144)
Lucius II (1144-1145)
Eugenius III (1145-1153)
Anastasius IV (1153-1154)
Adrian IV (1154-1159)
Alexander III (1159-1181)
Lucius III (1181-1185)
Urban III (1185-1187)
Gregory VIII (1187)
Clement III (1187-1191)
Celestine III (1191-1198)
Innocent III (1198-1216)
Honorius III (1216-1227)
Gregory IX (1227-1241)
Celestine IV (1241)
Innocent IV (1243-1254)
Alexander IV (1254-1261)
Urban IV (1261-1264)
Clement IV (1265-1268)
Gregory X (1271-1276)
Innocent V (1276)
Adrian V (1276)
John XXI (1276-1277)
Nicholas III (1277-1280)
Martin IV (1281-1285)
Honorius IV (1285-1287)
Nicholas IV (1288-1292)
Celestine V (1294)
Boniface VIII (1294-1303)
Benedict XI (1303-1304)
Clement V (1305-1314)
John XXII (1316-1334)
Benedict XII (1334-1342)
Clement VI (1342-1352)
Innocent VI (1352-1362)
Urban V (1362-1370)
Gregory XI (1370-1378)
Urban VI (1378-1389)
Boniface IX (1389-1404)
Innocent VII (1404-1406)
Gregory XII (1406-1415)
Martin V (1417-1431)
Eugenius IV (1431-1447)

Nicholas V (1447-1455)
Calixtus III (1455-1458)
Pius II (1458-1464)
Paul II (1464-1471)
Sixtus IV (1471-1484)
Innocent VIII (1484-1492)
Alexander VI (1492-1503)
Pius III (1503)
Julius II (1503-1513)
Leo X (1513-1521)
Adrian VI (1522-1523)
Clement VII (1523-1534)
Paul III (1534-1549)
Julius III (1550-1555)
Marcellus II (1555)
Paul IV (1555-1559)
Pius IV (1559-1565)
Pius V (1566-1572)
Gregory XIII (1572-1585)
Sixtus V (1585-1590)
Urban VII (1590)
Gregory XIV (1590-1591)
Innocent IX (1591)
Clement VIII (1592-1605)
Leo XI (1605)
Paul V (1605-1621)
Gregory XV (1621-1623)
Urban VIII (1623-1644)
Innocent X (1644-1655)
Alexander VII (1655-1667)
Clement IX (1667-1669)
Clement X (1670-1676)
Innocent XI (1676-1689)
Alexander VIII (1689-1691)
Innocent XII (1691-1700)
Clement XI (1700-1721)
Innocent XIII (1721-1724)
Benedict XIII (1724-1730)
Clement XII (1730-1740)
Benedict XIV (1740-1758)
Clement XIII (1758-1769)
Clement XIV (1769-1774)
Pius VI (1775-1799)

Pius VII (1800-1823)
Leo XII (1823-1829)
Pius VIII (1829-1830)
Gregory XVI (1831-1846)
Pius IX (1846-1878)
Leo XIII (1878-1903)
Pius X (1903-1914)
Benedict XV (1914-1922)

Pius XI (1922-1939)
Pius XII (1939-1958)
John XXIII (1958-1963)
Paul VI (1963-1978)
John Paul I (1978)
John Paul II (1978-2005)
Benedict XVI (2005-reigning)

Cyprian in his *Letter to all his People* [43 (40) 5], written in 251 AD, reminded his people that the faith of the pope is the faith of the Church:

They who have not peace themselves now offer peace to others. They who have withdrawn from the Church promise to lead back and to recall the lapsed to the Church. There is one God and one Christ, and one Church, and one Chair founded on Peter by the word of the Lord. It is not possible to set up another altar or for there to be another priesthood besides that one altar and that one priesthood. Whoever has gathered elsewhere is scattering.

The gates of hell shall not prevail against it!

Even though the Church is made up of sinners and saints (Mt. 5:13-16; 7:15-23; 10: 1-4; 13: 1-9, 24-50; 26:69-75; Mk. 3:19; Lk. 22:54-62; Jn. 6:70; 18:2-4) God promised that the gates of hell would never prevail against it (Mt. 16:18f) and he promised that he would keep it from error (Jn. 16:13; 1 Tm. 3:15). God has kept his promise.

He has entrusted the Church with a teaching office—the Magisterium—which consists of the bishops, the successors of the apostles, in union with the successor of Peter, the pope (Acts 8:30-31; 15:1-35; Eph. 2:20; 3:5; Jn. 14:16f).

The following is a short list of heresies that have attempted to prevail against the Church, but have failed.

Docetism (first century): This heresy denied Jesus' humanity.

Gnosticism (first century): Gnosticism is a mixture of Christianity, the oriental religions, and Greek philosophy. Salvation was only for the elect. All matter was evil; only the spirit was good. Marriage was to be avoided because it produced matter, children.

The Gnosticism of Basilides (second century): The following aspects made up this heresy: 1) Only the few are able to possess the true and secret knowledge (gnosis) that is necessary for salvation; 2) Only the soul is redeemed, the body corrupts; 3) Christians must reject Christ's crucifixion and only emphasize Jesus as the one sent by the Father.

The Gnosticism of Valentinus (second century): According to this heresy, the *aeon* Christ united himself with the man Christ to bring a secret knowledge (gnosis) to the elect, the Gnostics. This secret knowledge was directed at freeing the soul from the body so that the soul could enter into a spiritual realm after death.

The Gnosticism of Ptolemy (second century): The Law of the Old Testament is the product of a demiurge (the world-creator) who is neither the supreme God nor the devil. He is not perfect like the Supreme God nor is he the author of evil like the devil. This demiurge is known as the creator of the universe, the creator of matter which traps the soul in a body.

Marcionism (second century): This heresy denied the Old Testament's validity.

Montanism (second century): Montanism believed in an earthly thousand-year reign (Millenarianism) and taught mortal sins could not be forgiven. Many refused to marry because of a belief in the imminent return of Christ.

Modalism (second century): The Father, Son, and Holy Spirit were simply modes of one divine person.

Monarchianism (second century): Jesus was a human being who, at some point, perhaps at his baptism in the Jordan, received divine power from God.

Subordinationism (second century): Jesus was viewed as not being co-equal with the Father.

Sabellianism (third century): This heresy denied the distinction between the Father and the Son.

Patripassionism (third century): This belief argued that it was the Father, under the guise of the Son, who actually suffered and was crucified.

Novatianism (third century): Serious sinners could not be readmitted into the Church.

Manichaeism (third century): A religious and ethical doctrine which infiltrated much of Christian thought. It held that there were two equal eternal principles, one good, one evil, one spirit, one matter, one light, one darkness.

Arianism (fourth century): This heresy denied Jesus' divinity and the Trinity. This is the greatest heresy the Church has ever fought. Unlike Protestantism, which is still essentially a phenomenon of the western world, Arianism would affect and infect the entire Church.

Pneumatomachism (fourth century): This belief denied the divinity of the Holy Spirit and therefore of the Trinity.

Eunomianism (fourth century): A radical form of Arianism which denied the divinity of Christ and the Holy Spirit.

Donatism (fourth century): Donatism argued that grave sinners could not be readmitted into the Church, and the sacraments administered by those in mortal sin were to be held as invalid.

Priscillianism (fourth century): As a blend of Manicheanism, Docetism and Modalism, it denied the preexistence of the Son and denied the humanity of the Son.

Monophysitism (fifth century): This heresy claimed that Christ had only one nature.

Nestorianism (fifth century): This belief claimed that Jesus was two distinct persons and therefore denied the title "Theotokos," "God-bearer," "Mother of God."

Pelagianism (fifth century): This heresy denied the existence of "original sin," and believed that one could obtain salvation by works without grace or the Church.

Traducianism (fifth century): This heresy viewed the human soul as not created directly by God but generated by parents in the same way as a body.

Monothelitism (seventh century): Monothelitism denied that Christ had a human will and a divine will.

Paulicianism (seventh century): This heresy rejected the hierarchy of the Church and the sacraments of baptism, Eucharist, and marriage. They denied the Old Testament and parts of the New Testament.

Iconaclasm (eighth and ninth century): This heresy was led by the emperors that argued that icons fostered idolatry.

Albigensianism (eleventh century): This heresy rejected Church authority and the sacraments and denied the power of the civil authority to punish criminals. They viewed matter as evil. Suicide was considered the ultimate way of freeing oneself from one's evil body.

Catharianism (eleventh century): They renounced baptism and marriage. They viewed the body and matter as evil.

Waldensesism (twelfth century): Questioned the number of Church sacraments. They denied the validity of sacraments administered by an unworthy minister; they rejected purgatory and devotion to saints.

Lollardism (fourteenth century): Argued that the Bible should be in the language of the local people. They rejected the doctrine of Transubstantiation in favor of a simply spiritual presence in the Eucharist. They denied the role of the priest as a secondary mediator to the one mediator, Christ.

Hussitism (fifteenth century): Hussitism rejected the Sacrament of Penance, communion under one kind, and condemned the abuse of indulgences.

Protestantism (sixteenth century): Protestantism has taken on so many forms that it is hard to describe its belief system accurately for all. So for the purpose of this section of the book, I will describe the essential beliefs of Protestantism at the time of the Protestant Reformation.

Protestantism can primarily be summarized as follows: 1) Justification is by faith alone, *sola fides*; 2) The Bible alone is the rule of faith, *sola scriptura*; 3) "original sin" perverted human nature as opposed to wounding it; 4) Only baptism and the "Lord's Supper" are sacraments; 5) They rejected transubstantiation for consubstantiation or simple presence or symbol; 6) There is no need for a pope or bishops; 7) Mary's role in the Church is too great; 8) Indulgences should be rejected.

Gallicanism (seventeenth century): Held that a local church was autonomous and not answerable to the pope.

Jansenism (seventeenth century): Argued that one was without free will, that Christ did not die for all, that Christ's humanity was overemphasized, and that only the most holy were to receive the Eucharist.

Febronianism (eighteenth century): The state, guided by the Scriptures and subject to an ecumenical council, was to determine Church affairs. The pope was not to interfere in the affairs of the state.

Americanism (nineteenth century): Argued that there was a unique compatibility between Catholicism and American values. It argued that the United States held a providential role in guiding the universal Church into the modern age and particularly into the sphere of contemporary social issues.

Modernism (twentieth century): In its extreme form, it denied Christ's divinity, the sacredness of the Bible, the sacredness of the Church, and it believed that doctrines should change with the times; thereby denying infallible teachings.

Secular Christianity (twentieth/ twenty-first century): Secular Christianity is related to Modernism and is an outgrowth of Modernism. As opposed to molding and changing cultural values through the power of Christianity, this heresy seeks to mold and change Christianity according to modern cultural and secular values.

The Promise

The Church deals with heresies continuously, yet the promise of Jesus that he would be with the Church always (Mt. 28:20) and that the gates of hell would not prevail against it (Mt. 16:18f) has been kept.

Heresies have existed in the past and will continue into the future. In fact, heresies tend to transform themselves with every succeeding generation. Much of the "new age" movement can find its roots in many of these heresies. To accept any of these heresies is to say that God did not keep his promise of keeping the Church from error (Mt. 16:17-19; Jn. 16:13; 1 Tim. 3:15). How can that be?

Let us ponder once more the words of Ignatius of Antioch, the friend of the apostles John, Peter and Paul:

> *For as many as belong to God and Jesus Christ these are with the bishop. [A]s many as repent and come to the unity of the Church, these...shall be of God, to be living according to Jesus Christ. Be not deceived, my brethren, if any one follow a maker of schism, he does not inherit the Kingdom of God; if any man walk in strange doctrine he has no part in the Passion* (*Philadelphians*, 3, trans. Lake).

Jesus told Peter, *Kepa*, that he was the Rock, the *Kepa*, upon which he would build his Church: He promised that the gates of hell would not prevail against it (Mt. 16:18f). There was no such thing as Protestantism for 1500 years. The very word Protestant comes from the "protesting" of Catholicism. If we accept the Protestant claim that the Catholic Church went wrong during these 1500 years, then Jesus becomes a liar. Once more, Jesus said that he would build his Church on the Rock, on Peter, and the gates of hell would not prevail against it (Mt. 16:18f; 28:20). If Catholicism was wrong during any period during these 1500 years in its infallible teachings, or since those 1500 years, then the gates of hell would have had to prevail against the Church. Would Christ have allowed souls to be misled for 1500 years? Impossible! He has always been with his Church and will always be with her for all eternity (Mt. 28:20).

The major councils of the Church and the assurance of the true faith!
The councils of the Church that were approved and guided by the popes have protected the Church throughout the ages. For every series of heresies, God protected his Church through the successors of the apostles, the bishops, and the successor of the head of the apostles, Peter, the pope.

Council of Jerusalem (ca. 50): The apostles affirm the role of Gentiles in the Church.

Council of Nicaea (325): The council condemned the heresy of Arianism and affirmed that the Son was consubstantial (one with) the Father. The heresy of Arianism argued that the Son was created and not co-eternal with the Father. Arianism therefore denied the reality of the Trinity.

Constantinople I (381): The council condemned the heresy of Macedonianism which argued for a hierarchy in the Trinity instead of an equality. The council declared that the Holy Spirit was consubstantial (one with) the Father and the Son.

Councils of Hippo (393), Carthage III (397), and Carthage IV (419): A list of books are compiled that will become known as the Bible.

Council of Ephesus (431): Condemned Nestorianism and Pelagianism. The Heresy of Nestorianism denied the title "Mother of God" thus separating Christ's human nature from his divine nature; thereby making Christ essentially two distinct Persons. Pelagianism held essentially five key heretical points: 1) Adam would have died whether he sinned or not. 2) The sin of Adam injured only himself and was not passed on to further generations. 3) Newborn children are not affected or wounded by "original sin." 4) Christ's salvific event was not absolutely necessary for salvation, since sinless people existed prior to Christ. 5) One could work oneself into heaven by means of one's human efforts alone.

In response, the council affirmed the reality that the Son of God was the second Person of the Trinity and that he had two natures, one human, one divine—without change, confusion, separation or division between the natures. It thus affirmed the title of Mary as the "theotokos" "God-bearer," the "Mother of God." Against Pelagianism the Church affirmed the necessity of Christ's life, death, and resurrection for our salvation and the wiping away of "original sin." It affirmed that grace was necessary for salvation and that one could not earn or work oneself into heaven without the aid of grace. It affirmed that "original sin" is passed down to the entire human race and is cleansed in baptism.

Council of Chalcedon (451): The council condemned Monophysitism. Monophysitism denied Christ's two natures (divine and human) and argued for a composite nature. The council reaffirmed the teaching that the Son of God was one Person with two natures, without change, separation, confusion, or division between the natures. Jesus was fully human, fully divine. He was God and man.

Constantinople II (553): The council re-condemned the Nestorian heresy.

Constantinople III (680): The council condemned Monothelitism which argued that Christ had only one will. Constantinople affirmed that Jesus had two wills, a human will and a divine will. His human will was in perfect conformity with his divine will.

Nicaea II (787): The council condemned Iconoclasm which forbade the use of images as prayer aids. Nicaea affirmed the use of images for inspiring prayer. The incarnation, an icon of God, made images of the invisible God visible.

Lateran I (1123): It issued decrees banning simony (the buying and selling of something spiritual, such as religious offices) and lay investiture (the appointing of bishops by lay persons, as opposed to by the Church). It also affirmed the gift of celibacy in the priesthood.

Lateran II (1139): It ended the papal schism between Innocent II and Anacletus II. Anacletus was declared an anti-pope. Clerical celibacy was also reaffirmed and usury—the taking of interest for a loan—was prohibited.

Lateran III (1179): It condemned the Cathari who renounced baptism and marriage.

Lateran IV (1215): It condemned the Albigenses and Waldenses.

Albigensianism rejected the sacraments and Church authority. The Waldenses rejected the sacraments, purgatory, the communion of saints and Church authority. The council reaffirmed its always held beliefs in these teachings.

Lyons I (1245) deposed Frederick II and planned a crusade to free the Holy Land.

Lyons II (1274) reunited the Church with the Orthodox churches and enacted reforms in discipline regarding the clergy.

Vienne (1311-1312) enacted reforms in the Church and abolished the Knights Templars.

Constance (1414-1418) ended the papal schism and condemned the theology of John Huss.

Basle, Ferrara, Florence (1431-1445) reunited the Church with the Orthodox churches and again enacted disciplinary reforms.

Lateran V (1512-1517) dealt with the neo-Aristotelian influences in the Church and also enacted disciplinary reforms for the clergy.

The Council of Trent (1545-1563) affirmed what Protestantism denied. It reminded the Protestants of the beliefs that were always held from the beginning of the Church. It affirmed that the deposit of faith was found in Sacred Scripture and Sacred Tradition. It affirmed the reality of the seven sacraments. The doctrine of Transubstantiation was reiterated; that is, that bread and wine, once consecrated, become the body, blood, soul, and divinity of Christ. It declared that justification is by faith, but not by faith alone: works are necessary. It rejected the negative view held by Protestants regarding human nature. Protestants believed that "original sin" destroyed human nature; The Catholic Church would reassert the faith and declare that "original sin" "wounded" but did not destroy human nature. The Catholic Church would reassert the reality of free will and the reality of providence— early Protestantism believed in absolute predestination and the lack of free will; that is, some people are predestined to heaven and some to hell.

Vatican I (1869-1870) clarified and reaffirmed the always held teaching of papal primacy in honor, jurisdiction, and infallibility.

Vatican II (1962-1965) was a pastoral council that sought "renewal, modernization, and ecumenism." Vatican II also reasserted the faith of 2000 years of Catholicism.

Christ has always been with his Church and will always be with her for all eternity (Mt. 28:20). "You are Peter and upon you I will build my Church and the gates of hell shall not prevail against it" (Mt. 16:18f).

Why is there so much confusion in belief among Protestants?

Confusion in belief is the necessary consequence of having no apostolic successors—that is, authentic bishops—to resort to in determining questions of faith and morals. When we examine the main Protestant denominations we see that within each denomination, whether we are dealing with Baptists, Methodists, Presbyterians, etc., there are divergent beliefs, often radically divergent beliefs. This has led, unavoidably, to fundamentalism, the strict literal interpretation of the Scriptures.

Since the Scriptures can be interpreted in so many different ways, and since Protestantism does not have an infallible teaching office to determine correct interpretations, one is left with resorting to a narrow, literalistic interpretation of the Scriptures. This is the only way to assure a clear interpretation of the Gospel message, so it seems. But in reality, even among the strictest of fundamentalists, divergent beliefs abound.

Protestantism, although not conscious of it at the time of the Reformation, was doomed to failure from the start when it rejected the authority of the successors of the apostles.

Luther, the first Protestant, during a moment of depression, recognized the damage that he had begun when he saw denominations popping up in every place. During a moment of self-reflection, he wrote:

> *Who called you to do things such as no man ever before? You are not called.... Are you infallible?.... See how much evil arises from your doctrine.... Are you alone wise and are all others mistaken? Is it likely that so many centuries are wrong?.... It will not be well with you when you die. Go back, go back; submit, submit* (Grisar Hartman, *Luther,* St. Louis: Herder Book Co., 1914, II, 79; V, 319ff).

What gave rise to the birth of Protestantism?

Humanism

The development of the initial pangs of humanism brought to the fore the milieu for the birth of Protestantism. The sixteenth century exemplified a shift from the supernatural to the natural, from a God-centered world to a person-centered world. For Luther and Calvin a man chose his religion for himself, instead of accepting a religion revealed and dictated by God. The emphasis was no longer on religion changing a person as much as a person changing religion.

The outcome of this line of thinking is the development of religious truths based on the thoughts and personal inspirations of varying individuals—thus 33, 000 mainline denominations.

Scholastic Deterioration

Scholasticism which had always been at the heart of Catholic theology was now viewed as passé. Nominalism was now in vogue, and this Nominalism would have a profound impact on the birth of Protestantism.

Scholasticism emphasized the relationship between faith and reason. For Scholastics faith and reason could never contradict.

Nominalism however was more subjective. The human mind was subject to distortion. Faith and reason could contradict each other. The power of reason was subject to flaws—hence, the Protestant doctrine of *sola fides*, by faith alone. Since one cannot depend on reason, one must depend on faith alone.

Pseudo-Mysticism

The spirituality of the time was marred by skepticism and a discomfort with reason. A strict understanding of "original sin" and concupiscence would infect many. For many "original sin" had completely destroyed human nature—free will, reason, and the ability to do good was thus extinguished. The core of Protestantism could be found in the thoughts of pseudo-mysticism, predestination and faith alone.

In Catholic spirituality, by contrast, "original sin" had only wounded human nature not destroyed it, leaving the person basically good. A person's free will and his or her ability to reason was not destroyed, albeit inclined to error and sin.

Clerical Laxity

The two hundred and fourteenth pope, Alexander VI, was a weak and immoral man. While he protected the faith infallibly, he was subject to many human weaknesses. He fathered several illegitimate children and thus betrayed his vow of chastity; he sought after wealth and denied his vow of poverty; he engaged in bribery, nepotism, simony, and the misuse of indulgences. He also was associated with mysterious deaths. He died by poisoning.

The bitter legacy of Alexander VI's sinfulness would be responsible in great part for the heresy of Protestantism some fourteen years later.

The clergy of the time had fallen to its lowest level in Christian history. Priests were enmeshed in secular humanism and secular pursuits. They disregarded their clerical clothing, abandoned their prayers and their concern for their parishioners. They openly rejected clerical celibacy and sought after wealth. It was not unusual to see Cardinals at the age of thirteen.

Between 1049 and 1274 the Church had seventy-four canonized bishops. Between 1378 and 1521 the Church had four. The Church had fallen to its most severe level of debauchery.

Abuse of Indulgences

The problem with indulgences, prior to the Protestant Reformation, was not so much a problem of dealing with the belief as much as a problem of dealing with the abuse of the use of indulgences. Sinners often paid professional penitents to expiate their temporal punishments; some priests sold indulgences without pointing out the requirements that are associated with indulgences; and some priests sold indulgences to raise money to build churches. The right motives and the right understanding of the role of indulgences were not present in these abuses.

Abusive priests would say, "Before the coin hits the bottom of the can, the soul of a loved one in purgatory will be freed." Abuse was rampant!

The Rise of Secular Power

The moral failure of the clergy gave the secular authorities the perfect opportunity to assert themselves. The Church's moral authority and influence over the people had diminished to such an extent that Church censures, excommunications, and pronouncements were given no more than "lip service." The days of a King dressed in sackcloth begging and groping for forgiveness on the ground of a papal palace was over. The dawn of the separation of Church and State was approaching.

It is no surprise that the first Protestants were protected by states. Germany protected Luther, Henry VIII made himself the head of the Church, and Calvin and Zwingli sought refuge from friendly kings and princes seeking freedom from the influence of the popes.

The Printing Press

Prior to the sixteenth century books were found primarily in the Church and amongst the elite. The cost and time to handwrite (most often by monks) a book was exhaustive and consequently too expensive for the ordinary person to own. Furthermore, the majority of the populace was illiterate.

By the time of Luther, all this would change. The printing press had been invented. What took years to do, now took days, weeks, or months. A tract that once took several months to produce in limited amounts could now be produced quickly and in essentially unlimited amounts.

Because of the printing press, the spread of Luther's revolt and theology would extend throughout Christendom almost overnight.

Harshness of the Times

Luther, in all likelihood, never wanted to leave the Church he served as an Augustinian Catholic monk. Unfortunately these were harsh times. They were not times of dialogue, but of confrontation, not of reconciliation, but of destruction, not of reform, but of keeping the *status quo*.

Luther, sadly and unlike those who would follow, was pushed away from the faith and out of the Church.

History and Christendom would pay a heavy price for this.

III
SACRAMENTS

Are sacraments just symbols?

For the Catholic sacraments are efficacious. That is, they produce what they signify. A sacrament imparts grace to the individual (Acts 2:38; 8:17; 19:4-7; 1 Pet. 3:19-22).

The seven sacraments of the Church are foreshadowed in a powerful manner in the Hebrew Scriptures, in the Old Testament signs and symbols of the covenant—circumcision, anointing, consecration, the laying on of hands, sacrifice, Passover. These signs of the Old Covenant foreshadow the signs of the New Covenant—Baptism, Confirmation, Holy Orders, Eucharist, Penance, Anointing of the Sick and Matrimony. Each of the sacraments of the "new covenant," of the New Testament, can find similarities in one or more of the Old Covenant, Old Testament, signs and symbols—baptism replacing circumcision, anointing of the sick replacing the ancient anointings, consecrating or ordaining priests by the laying on of hands replacing the consecration of kings and Levitical priests, the institution of the Eucharist as presence and sacrifice replacing the Old Testament sacrifices of lambs and the Passover experience.

What do Catholics mean by being "born again" (Jn. 3:3-5) and why do they baptize children?

Baptism is God's most beautiful and magnificent gift.... We call it gift, grace, anointing, enlightenment, garment of immortality, bath of rebirth, seal, and most precious gift. It is called gift because it is conferred on those who bring nothing of their own; grace since it is given even to the guilty; Baptism because sin is buried in the water; anointing for it is priestly and royal as are those who are anointed; enlightenment because it radiates light; clothing since it veils our shame; bath because it washes; and seal as it is our guard and the sign of God's Lordship.

Augustine of Hippo, Letter to Jerome
(Oratio 40, 3-4: PG 36, 361C)

In the Old Testament, the reality of baptism is beautifully prefigured in Ezekiel 36:25-27:

I will sprinkle clean water upon you to cleanse you from all your impurities, and from all your idols I will cleanse you. I will give you a new heart and place a new spirit within you, taking from your bodies your stony hearts. I will put my spirit within you and make you live by my statutes, careful to observe my decrees.

In 1 Peter 3:20-21 we read:

[Eight persons, in the account of Noah's flood] were saved through water. This prefigured baptism, which saves you now.

Jesus taught Nicodemus that one must be born again by *water* and the *Spirit* (Jn. 3:5), not by the Spirit *only*, but by *water* and the *Spirit*.

Most fundamentalists view baptism as merely a symbol which signifies something that has already taken place. That is, one becomes "born-again" by accepting Jesus as one's Lord and Savior (Jn. 3:16), and then one is baptized to symbolically affirm what has already taken place. Baptism, for evangelicals, therefore implies the use of reason. Consequently, children under the age of reason cannot make such a reasoned decision. Children should not be baptized according to their view.

Besides the idea of infant baptism being contrary to their theology, fundamentalists argue that there is no proof of infant baptism in the Bible and since there is no proof of infant baptism in the Bible, it is not a Christian practice.

For the Catholic to be baptized is what it means to be "born again." Baptism is a sacrament with real power and it is a sacrament which is necessary for salvation, for it is by baptism that we are "born again" of water and the Spirit (Jn. 3:5; Mk. 16:16). God "saved us through the bath of rebirth and renewal by the holy Spirit, whom he richly poured out on us through Jesus Christ our Savior, that we might be justified by his grace and become heirs in hope of eternal life" (Titus 3:5-7). In baptism one enters into Christ's death and resurrection (cf. Rom. 6:3-4). One has put on Christ in baptism (Gal. 3:27). Baptism cleanses one from "original sin," personal sin, and the punishment for sin (Mk.16:16; Jn. 3:5; Acts 2:38f; 22:16; Rom. 6:3-6; Gal. 3:7; 1 Cor. 6:11; Eph. 5:26; Col. 2:12-14; Heb. 10:22).

Psalm 51:7 states: "In guilt was I born, and in sin my mother conceived me." One becomes a new creation in Christ and a partaker in the divine nature (2 Pet. 1:4). One becomes a member of the Church as an adopted child of God (cf. 1 Cor. 12-13, 27). One becomes a Temple of the Holy Spirit (Acts 2:38; 19:5f) with an indelible mark or character on the soul which enables one to share in the priesthood of Christ and in his passion (Mk. 10:38f; Lk. 12:50).

Because of this reality, infants are encouraged to be baptized. As Irenaeus explains in *Against Heresies* (cf. 2, 22, 4) (ca. 180): "Jesus came to save all for all are reborn through him in baptism—infants, children, youths, and old men." To deny a child baptism is to deny a child the precious gifts of baptism. How contrary to God's will (Mt: 19:14; Lk. 18:15-17): "Let the children come to me, sayeth the Lord."

Saint Augustine of Hippo (ca. 415) so poignantly states in his *Letter to Jerome* (166, 7, 21) the importance of the baptism of children:

Anyone who would say that even infants who pass from this life without the participation in the Sacrament [whether by a baptism of desire, blood, or water] shall be made alive in Christ truly goes counter to the preaching of the apostles and condemns the whole Church, where there is great haste in baptizing infants because it is believed without doubt that there is no other way at all in which they can be made alive in Christ.

St. Hippolytus of Rome (ca. 215) argues: "Baptize first the children; and if they can speak for themselves, let them do so. Otherwise, let their parents or other relatives speak for them" (*Apostolic Tradition*, 21). The ecclesiastical writer Origen in 244 AD wrote: "The Church received from the apostles the tradition of giving baptism to infants" (*Commentary on Romans*, 5:9).

In terms of Bible quotations with reference to infants in particular, I would refer you to the following quotes, Acts 16:15, 33 and 1 Corinthians 1:16. In these quotations we see that whole families were baptized. Given the culture of the ancient world, this most likely implied the baptism of infants. How can whole households not have any infants, any children?

In the Acts 2:38-39 we read a direct account of where Peter baptized adults and children: "Peter said to them, 'Repent, and be baptized every one of you in the name of Jesus Christ for the forgiveness of your sins; and you shall receive the gift of the Holy Spirit. For the promise is to you and to your *children* and to all that are far off, every one whom the Lord our God calls to him.... (RSV)."

Many argue that one must be old enough to accept the faith: No one, not even your parents, can stand in for you. This might sound appealing but it is contrary to the Scriptures. God often bestowed spiritual gifts on peoples because of the faith of others. The centurion's faith brought about the healing of his servant (Mt. 8:5-13); the Canaanite woman's faith brought about the healing of her daughter (Mt. 15:21-28); and in Luke 5:17-26 a crippled man is healed by the persistent faith of his friends. A parent's faith bestows the gifts of baptism upon their children.

Furthermore, Paul in Colossians 2:11-12 reminds us that baptism replaces circumcision for the Christian. In the Old Covenant, the Old Testament, one became a member of the people of God through circumcision on the eighth day. Now in the New Covenant, the New Testament, one becomes a member of the people of God through baptism as early as possible! For the Christian baptism is a replacement for circumcision (Col. 2:11-12). If Jewish parents would covenant with God on behalf of their eight-day old children through the command to circumcise their children, then Christian parents covenant with God on behalf of their children through the command to baptize. How could people deny children entrance into the covenant, into the people of God? As Jesus said: "Let the children come to me for the Kingdom of God belongs to such as these" (cf. Mk. 10:14; Lk. 18:15-16).

And as alluded to above, the fact that "whole households" were baptized in the New Covenant makes absolute sense since "whole households" were circumcised in the Old Covenant (Gen. 17:12-14), including house-born slaves and "foreigners acquired with money." Children were circumcised under the Old Covenant, and under the New Covenant they were baptized.

In Judaism a child had no say as to whether he was circumcised or not! On the eighth day he became a part of the people of God by virtue of the will of his parents and the act of circumcision. The same applies with the baptism of children! And just as in the Old Testament, when one reached the age of reason and could reject the gift received as an infant, one could likewise, in the New Testament times, reject the gift of baptism. The gift is given to be affirmed or rejected, to be nourished or to be allowed to die.

The following quotes are also worth noting when understood within the context of all the previous quotes from this section on infant baptism (Mt. 18:14; 19:13f; Mk. 10:13-16; Lk. 18:15-17). To deny infant baptism is to thwart Jesus' call to the children when he said, "Let the children come to me. The Kingdom of God is for such as these" (Mt. 19:14).

Hermas, often referred to as the man named by Paul in Romans 16:14, writes in *The Shepherd, Parable 9*: "One cannot enter the kingdom of God without coming up through the water of baptism so that one may attain life.... The person goes down into the water dead, and then comes up alive." In the *Epistle of Barnabas* (11; 16) (ca. 96), the Barnabas often referred to as the companion of the apostle Paul by men such as Clement of Alexandria, Eusebius, and Jerome, we read: "In baptism we receive the forgiveness of our sins, and trusting in the name of the Lord, we become a new creation.... God truly dwells in our house, in us."

Given these early teachings and the teachings of the Bible, why do so many reject the efficaciousness of baptism and the baptism of infants?

In the ancient *Jerusalem Catechesis* we find a precious synthesis on the beauty of baptism:

> *As our Savior spent three days and three nights in the depths of the earth, so your first rising from the water represented the first day and your first immersion represented the first night. At night a man cannot see but in the day he walks in the light. So when you were immersed in the water it was like night for you and you could not see, but when you rose again it was like coming into broad daylight. In the same instant you died and were born again; the saving water*

was both your tomb and your mother.... Let no one imagine that baptism consists only in the forgiveness of sins and in the grace of adoption. Our baptism is not like the baptism of John, which conferred only the forgiveness of sins. We know perfectly well that baptism, besides washing away our sins and bringing us the gift of the Holy Spirit, is...[an entrance into] the sufferings of Christ. That is why Paul exclaims: 'Do you not know that when we were baptized into Christ Jesus we were, by that very action, sharing in his death?' By baptism we went with him into the tomb (Cat. 21, Mystagogica 3, 1-3: PG 33, 1087-1091 quoted from *The Liturgy of the Hours*).

- The Anabaptists, the forerunners of the Baptists, denied infant baptism in the sixteenth century. It is significant that in an unusual moment of unity among the Protestant denominations the Anabaptists were condemned as denying a long held practice of Christianity.

Baptism by blood and desire for adults and infants

The Scriptures and the Tradition of the Church refer to three forms of baptism for salvation. One is the one that we are all familiar with where Jesus sends all his disciples throughout the world to baptize in the name of the Father, and of the Son, and of the Holy Spirit (Mt. 28:19-20).

The other two forms are referred to as baptism by blood and baptism by desire. The Church has always maintained that those who suffer death for the sake of the faith before having been baptized are baptized by their blood, by their death in, for, and with Christ. The fruits of the sacrament of baptism are given to the person even though they did not receive the sacrament. Similarly, those who die before being baptized and yet desired baptism in their lifetime likewise receive the fruits of the sacrament without receiving the sacrament itself. Those who are moved by grace and may not be explicit Christians are equally considered as having a baptism by desire. And children who die before baptism are baptized by the desire of the parents or the mystical body, the Church.

The Scriptures point to the salvation of the Holy Innocents by Herod (cf. Mt. 2:16-18): The infants that were massacred died for Christ and therefore can be considered to have been baptized by their blood. On the cross of Calvary, the good thief, Dismas, called for mercy and received God's forgiveness and salvation. He certainly could not come down off the cross to be saved in a water baptism. He was saved and baptized by his desire (Lk. 23:42-43).

(For a more detailed understanding of the baptism of desire and the baptism of blood and its relationship to grace, go to the section referring to the salvation of "non-explicit Christians").

Does baptism require immersion?

From ancient times baptism took place by a triple immersion in water with the use of the Trinitarian formula (i.e., I baptize you in the name of the Father, and of the Son, and of the Holy Spirit). But the early Church also recognized as valid the pouring of water over the head of a person with the same use of the Trinitarian formula. The *Didache* (ca. 65), often attributed to the apostles, states:

> *Regarding baptism, baptize this way: Baptize in the name of the Father, and of the Son, and of the Holy Spirit in running water. But if you have no running water, baptize in any other. [In that case] pour [water] three times on the head in the name of the Father, and of the Son, and of the Holy Spirit* (*Didache, 7*, trans. Francis Glimm in *The Fathers of the Church: A New Translation,* Washington: The Catholic University Press, 1962).

The early Church theologians such as Justin Martyr (ca. 148-155), Tertullian (ca. 206), and Hippolytus (ca. 215) emphasized that baptism could be done by immersion, infusion, or aspersion.

- The *Didache* is one of the oldest existing Christian documents, predating many of the New Testament writings. It is believed to have been written by the twelve apostles as a guide for Gentile Christians in terms of Church law and order. Some scholars, however, like to place the writing of the *Didache* shortly after the death of the last apostle John (ca. 100) at around 120 AD. In either case, the *Didache* was considered by men like Hermas, Irenaeus, Origen, Clement of Alexandria, Eusebius, Cyprian, Lactantius, and many others as part of Sacred Scripture. The *Jerusalem Codex* of the New Testament contains the *Didache* within it. Again, it is the pope and the bishops in union with him that excluded the *Didache* from the final canon in the fourth century.

Baptism of the dead?

Mormons like to quote 1 Corinthians 15:29 to affirm the practice of baptizing the unbaptized dead. This is a practice that is approximately 150 years old and is the invention of the founder of Mormonism, Joseph Smith. It is not mentioned in the writings of the early Church—not even once. And the Scriptural reference to Corinthians is never affirmed by the apostles in the Scriptures or anywhere else. Ironically, even the *Book of Mormon*, which supposedly contains the fullness of the Gospel, does not mention!

The baptism of the dead has never been practiced by informed and orthodox Christians. Mormons are the first to make it a practice. The

Scriptures are clear that one acquires one's salvation or ends up being condemned during this lifetime, not in the afterlife.

But how do we explain 1 Corinthians 15:29? Paul was making what is philosophically called an *Ad Hominem* argument for the resurrection of the body.

Just north of Corinth was the city named Eleusis. Pagans in that city adopted the practice of baptizing themselves for the dead. Homer, a Greek pagan, alludes to this practice in his *Hymn to Demeter*. In other pagan cities people practiced the baptism of corpses. Being a large bustling port city, it would not be unthinkable that some Corinthian Christians would have incorporated into their practice—erroneously—the Eleusian observance of baptizing themselves for the dead.

Influenced by Greek culture, the Corinthians viewed the soul as somehow trapped within the body. The materialism of the world was somewhat unappealing to Greek sentiments. The Corinthian Christians therefore found the resurrection of the body as somewhat unattractive or unbelievable. The logic was flawed: They baptized themselves for the dead, yet did not believe in the resurrection!

Paul takes advantage of this pagan practice of baptizing oneself for the dead—which was adopted by this group of Corinthian Christians—by arguing for the resurrection of the body. Paul basically argues that it is absurd to baptize oneself for the dead if one does not believe in the bodily resurrection.

Paul's whole theology of baptism makes it quite clear that Paul would never have approved of this practice. Paul was simply using this practice to his advantage.

Baptism is necessary for salvation, and so one can see why the Mormons would want to baptize the dead. However, salvation or condemnation takes place in this earthly journey, not in the afterlife. Hell is eternal (Mt. 25:41; 2 Thess. 1:6-9) as is heaven (Mt. 5:8; 25:33-40; Rom. 8:17; Phil. 4:3; Heb. 12:23; Rev. 3:5; Mt. 5:12; Jn. 12:26; 14:3; 17:24; 1 Cor. 13:12, etc.). Why then would one baptize the dead? For what reason?

Even the book of Mormon, ironically, seems to contradict the belief in the baptism of the dead (i.e., Alma 34:31-35; 5:28, 31; 2 Nephi 9:38; Mosiah 16:5, 11; 26:25-27). Only in this life can one gain or lose one's salvation!

Because one can interpret the Scriptures in a legion of ways, one is subject to error. That is why there are over 33,000 mainline Protestant denominations and 150,000 pseudo-Christian denominations. That is why the Church that put the Bible together is the only one that can interpret it properly and infallibly—and who put the Bible together?
The Catholic Church!

Where do we find the Sacrament of Confirmation?

After coming from the place of baptism we are thoroughly anointed with a blessed unction. After this, the hand is imposed for a blessing, invoking and inviting the Holy Spirit. The unction runs on the body and profits us spiritually, in the same way that baptism is itself a corporal act by which we are plunged into water, while its effect is spiritual, in that we are freed from sins....
Tertullian (ca. 200)
Baptism (7, 1, 2, 5; 8, 1)

Confirmation perfects baptismal grace. Notice that in Acts 19:5-7 Paul "lays his hands" on the recently baptized invoking the Holy Spirit, thereby confirming them. Likewise, in Acts 8:14-17 Peter and John "lay their hands" on the converts of Samaria, for as the Bible says: "the Spirit had not come upon any of them; they had only been baptized in the name of the Lord Jesus" (Acts 8:16). Peter and John were confirming, perfecting, what had begun at baptism in the converts of Samaria.

Once confirmed we are strengthened by the Holy Spirit to be powerful witnesses of Christ's self-communicating love to the world. We become strengthened members in the mission of the Church, the proclamation of the Gospel. Like baptism, a sacred mark or seal is imprinted on the soul, forever changing it (cf. 2 Cor. 1:21-22; Eph. 1:13). In Acts 1:6-8 we see how, despite being baptized previously, the apostles received the gift of the Holy Spirit to be witnesses to the world.

In receiving this sacrament by a bishop or a delegated priest, one is making a commitment to profess the faith and to serve the world in word and deed as a disciple of Christ (Acts 19:5-6; 8:16-17; Heb. 6:1-2; 2 Cor. 1:21-22; Eph. 1:13).

Cyril of Jerusalem (350 AD) beautifully summarizes the power and the necessity of the Sacrament of Confirmation:

And to you in like manner, after you had come up from the pool of the sacred streams, there was given chrism, and this is the Holy Spirit (21 [3] 1). But beware of supposing that this is ordinary ointment. For just as the bread of the Eucharist after the invocation of the Holy Spirit is no longer simple bread, but the Body of Christ, so also this holy ointment is no longer plain ointment, nor, so to speak, common, after the invocation. Rather, it is the gracious gift of Christ; and it is made fit for the imparting of his godhead by the coming of the Holy Spirit. This ointment is applied to your forehead and to your other senses; and while your body is anointed with the visible ointment, your soul is sanctified by the Holy and life-creating

Spirit (21 [3] 3). Just as Christ, after his baptism and the coming upon him of the Holy Spirit went forth and defeated the adversary, so also with you; after holy baptism and the mystical chrism of the [Sacrament of Confirmation] and the putting on of the panoply of the Holy Spirit, you are able to withstand the power of the adversary and defeat him by saying, 'I am able to do all things in Christ who strengthens me' (21 [3] 4) (Mystagogic).

Why do Catholics believe the Eucharist is the Body and Blood of Jesus?

If the words of Elijah had power even to bring down fire from heaven, will not the words of Christ have power to change the natures of the elements [of bread and wine into the body and blood of Jesus]?
...The Lord Jesus himself declares: This is my body. Before the blessing contained in these words a different thing is named [bread]; after the consecration a body is indicated. He himself speaks of his blood. Before the consecration something else is spoken of [wine]; after the consecration blood is designated.

Ambrose (397d), On the Mysteries
(Cf. Nm. 52-54, 58: SC 25 bis, 186-188, 190)

Jesus came to us in Bethlehem, which means "house of bread," and was placed in a "manger" which is an eating vessel. Today Jesus is present in the tabernacle, those houses where the bread of eternal life, the body, blood, soul and divinity of Christ is present throughout the world. He is given to us in a manger, an eating vessel, the chalice and the paten at every Mass. From the very moment of his incarnation, his entrance into the world, the Son of God was pointing to his wonderful gift of the Eucharist!

Belief in the Eucharist as the Body and Blood of Christ has declined in recent years. This is the sad consequence of the growth of secularism, modernism, and fundamentalism.

Despite this, the Scriptures and Tradition affirm the Catholic position on the Eucharist. Let us examine a powerful passage from the Bible that supports the Catholic view:

The Jews...disputed among themselves, saying, "How can this man give us his flesh to eat?" So Jesus said to them, "Truly, truly, I say to you, unless you eat the flesh of the Son of man and drink his blood, you have no life in you; he who eats my flesh and drinks my blood has eternal life, and I will raise him up at the last day. For my flesh is food indeed and my blood is drink indeed. He who eats my flesh and drinks my blood abides in me, and I in him. As the living Father sent me, and I live because of the Father, so he who eats me will live because of me. This is the bread that came down from heaven, not such as the fathers ate and died; he who eats this bread will live forever.... Many of his disciples, when they heard it, said, "This is a hard saying; who can listen to it?" ...After this many of his disciples drew back and no longer went about with him (Jn. 6:52-58, 60, 66 RSV; also make reference to the following passages: Mt. 26:26-28; Mk. 14:22-24; Lk. 22:19f; 1 Cor. 10:16; 11:24f, 27, 29).

How can anyone deny the Real Presence? John's Gospel emphasizes not just the "body," but the "flesh" of Christ that one is called to partake in. Furthermore, when one looks at the word that John uses for "eat," the word that John uses is not the classical Greek word for eat, rather it is—during this particular time in history—a vulgar term used to describe animals eating. The best translation today would probably be "munch" or "gnaw." Obviously John is emphasizing the reality of the Body and Blood of Christ.

Some like to emphasize the fact that Jesus called himself a door (Jn. 10:9), a vine (Jn. 15:1), a lamb (Jn. 1:29), a light (Jn. 8:12), living water (Jn. 4:14), etc. They would claim that Jesus was clearly being symbolic here. They would argue consequently that Jesus in referring to the Eucharist as his Body and Blood was doing no different than calling himself living water.

While it is true there is a symbolic dimension to calling Jesus the Bread of Life, the early Christians, however, understood that Jesus was referring to that which transcended the simply symbolic. Jesus was talking about his Real Presence in the Eucharist, his real Body and Blood.

When we look to history, we recognize that Christians were often sent to their deaths by the Romans under the accusation of being cannibals—as testified to by the first century pagan historian Tacitus in his *Annals*. Where would they get such a thought unless the Real Presence was not obviously and fervently believed? (What Tacitus did not realize is that Catholics eat the Risen Christ! Cannibals eat dead flesh).

No one went to his or her death for proclaiming Jesus as a door, a vine, a lamb, or symbolic bread. No one went to his or her death in the amphitheater for worshiping a door or living water. The early Church always distinguished that which was symbolic from that which was to be taken literally. Acts 10:39 describes Jesus as being hung on a tree; all Christians knew he had been crucified on a cross; all Christians knew that Paul was using symbolic language.

It is also interesting to read that after Jesus' discourse on the Bread of Life, many of the disciples abandoned him. *Why would they abandon him if all they thought was that Jesus was using symbolic language?* Why run away if all that is being talked about is a symbol?

The disciples knew that Jesus was not simply talking symbolically. That is why they ran away! Genesis 9:3-4 and Leviticus 17:14 strongly forbade the eating or drinking of blood. The disciples abandoned Jesus because they knew of these quotes from the Old Testament and they knew that Jesus was talking literally.

The disciples never abandoned Jesus for being the "door," the "vine," the "lamb," or "the Son of God." They never abandoned Jesus when he said he was the way, the truth, and the life, and that no one goes to God the Father except through him (Jn. 14:6). Nor did they abandon him for forgiving sins, which only God can do! But they certainly ran away when they heard Jesus describe his Real Presence in the Eucharist.

Whenever in the Scriptures there was confusion among his disciples, Jesus corrected the misunderstanding and explained to them the true meaning of what he meant (see Jn. 3:3-5; Jn. 11:11-14; Mt. 19:24-26; Jn. 8:21-23; Jn. 8:31-36; Jn. 6:32-35). And when Jesus wanted his words to be taken literally he repeated and reaffirmed what he said (see Mt. 9:2-6; Jn. 8:56-59; Jn. 6:41-51).

Jesus, in this Eucharistic passage, not only does not explain away what he says, he re-enforces the literal meaning of what he is saying by repeating it over and over again. In fact, Jesus repeats six times in six verses the same literal truth (Jn. 6:53-58). Jesus wants to make the point perfectly clear. Notice, that Jesus does not run after the departing disciples and say, "Wait a minute, you misunderstood what I said! I was only talking symbolically!" Rather, he turned to the remaining disciples and said, "Are you going to leave me too" (Jn. 6:67)?

Now some like to point to verse 64 where we read, "It is the spirit that gives life, while the flesh is of no avail." They claim this proves that Jesus was only talking symbolically since flesh is worthless.

This argument makes no philosophical sense. What do they make Jesus out to be? Would Jesus at one moment be saying "eat my flesh" and then in the next moment be saying, "but my flesh is no good?" That would simply be absurd!

John 6:64 must be understood as John 3:6 is understood; that is, only by a gift of God can one truly comprehend and believe in what Jesus has said. Only by a gift from above, the gift of the Spirit, can one believe in the Real Presence of Christ in the Eucharist, for "no one can come to [Jesus] unless it is granted him by my Father" (Jn. 6:65).

Now one may argue, "How can Jesus be present in different ways?" That should be easy to answer for all Christians. Christ is present in the individual, in the congregation, in the minister, in the proclamation of the Word and so on. So too, Christ is sacramentally present in the Eucharist.

Let us examine one more powerful passage:

[A]nyone who eats the bread and drinks the cup of the Lord unworthily is answerable for the body and blood of the Lord. Everyone is to examine himself and only then eat of the bread or drink from the cup, because a person who eats and drinks without recognizing the body is eating and drinking his own condemnation (1 Cor.11:27-29, NJB).

Can a symbol bring one's own condemnation? Paul gets right to the point. We are dealing with the Real Presence of Christ in the Eucharist under the appearance, or what is technically called the "species," of bread and wine.

Let us now look at one who walked and talked and learned from the apostles themselves. What did he believe about the Eucharist? Would you believe the word of a person who lived in the sixteenth century who had no personal contact with an apostle or would you prefer the word of one who was taught by an apostle? I suspect that we would all prefer the testimony of a person that learned his Christianity from one of the apostles.

Ignatius of Antioch (ca. 107), the disciple of John and friend of Peter and Paul, writing only seven years after the death of the apostle John, reprimands the Docetists in his letter to the *Smyrneans* (6:7) for failing to believe in Christ's Real Presence in the Eucharist:

For let nobody be under any delusion; there is judgment in store for the hosts of heaven, even the very angels in glory, the visible and invisible powers themselves, if they have no faith in the blood of Christ. Let him who can, absorb this truth.... But look at the men who have those perverted notions about the grace of Jesus Christ which has come down to us, and see how contrary to the mind of God they are. They even absent themselves from the Eucharist and from prayer because they do not confess that the Eucharist is the flesh of our Savior Jesus Christ....

In another passage, Ignatius reminds us that there is only one authentic Eucharist:

Be careful...to use one Eucharist, for there is one flesh of our Lord Jesus Christ, and one cup for union with his blood, one altar, as there is one bishop with the presbytery [priesthood] and the deacons my fellow servants, in order that whatever you do you may do it according to God (Philadelphia, 4, trans. Lake).

In another passage to the Romans, Ignatius writes:

I have no taste for corruptible food nor for the pleasure of this life. I desire the Bread of God, which is the Flesh of Jesus Christ, who was of the seed of David; and for drink I desire His Blood, which is love incorruptible (Romans, 7:3, Jurgens).

Ignatius of Antioch is a giant in Christendom. His words were recognized as truth, for they came from the mouth of a disciple of the apostles John, Peter, and Paul.

Irenaeus, the friend of Polycarp, who in turn was the friend of the apostle John wrote:

Jesus declared the cup, a part of creation, to be His own Blood, from which He causes our blood to flow; and the bread, a part of creation, He has established as His own Body, from which He gives increase to our bodies (Against Heresies 5, 2, 2).

Justin Martyr, well-known by the disciples of the apostle John, wrote:

We call this food Eucharist...since Jesus Christ our Savior was made incarnate by the word of God and had both flesh and blood for our salvation, so too, as we have been taught, the food which has been made into the Eucharist by the Eucharistic prayer set down by Him, and by the change of which our blood and flesh is nourished, is both the flesh and the blood of that incarnated Jesus.

Justin further goes on to say:

None is allowed to share in the Eucharist unless he believes the things which we teach are true...for we do not receive the Eucharist as ordinary bread and ordinary wine, but as Jesus Christ our Savior.

Cyril of Jerusalem, another man acquainted with the disciples of John, wrote:

[Jesus] himself...having declared and said of the Bread, "This is My Body," who will dare any longer to doubt? And when He Himself has affirmed and said, "This is my Blood," who can ever hesitate and say it is not His Blood (Catechetical Lectures 22, Mystagogic 4).

Do not, therefore, regard the bread and wine as simply that, for they are, according to the Master's declaration, the Body and Blood of Christ. Even though the senses suggest to you the other, let faith make you firm. Do not judge in this matter by taste, but be fully assured by faith, not doubting that you have been deemed worthy of the Body and Blood of Christ (Ibid.).

The Real Presence of Christ in the Eucharist has always been held. No one seriously or significantly questioned the Real Presence of Christ until the eleventh century with the writings of Berengarius of Tours. Would God allow eleven centuries to go by with a false belief? If he said he would be with the Church for all eternity (Mt. 16:18; 28:20), then he would not lead it into error.

How sad it must be for those who do not receive the real Eucharist for it is, as Ignatius says in *Ephesians* 20, "the medicine of immortality, the antidote that we shall not die, but live forever in Jesus Christ."

Before we move on to the nature of the Mass whereby bread and wine become the Body and Blood of Christ under the appearance or "species" of bread and wine, I would like to leave you with a reflection on a very significant passage from the Scriptures, Luke 24:13-35. After the resurrection, Jesus in his glorified body joins two discouraged disciples on the way to Emmaus. Because of Jesus' glorified body, the disciples do not recognize Jesus until significantly "the breaking of the bread." Notice the similarity between Jesus' words at the Last Supper (Lk. 22:19) and his words in Luke 24:30-31: "[W]hile he was with them at table, he took bread, said a blessing [this implies a change], broke it, and gave it to them. With that their eyes were opened and they recognized him, but he vanished from their sight." Jesus vanished from their sight, but his presence was recognized in the "breaking of the bread." To the disciples, Jesus was "made known in the breaking of the bread" (Lk. 24:35).

Today, in every Catholic Church, Jesus is made known to us in the "breaking of the bread," in his body and blood, in his Real Presence under the "species" of bread and wine. And let us not take this gift lightly. For as Origen (ca. 185-253) states in his homily on Exodus:

You are accustomed to take part in the divine mysteries, so you know how, when you have received the Body of the Lord, you reverently exercise every care lest a particle of it fall, and lest anything of the consecrated gift perish. You account yourself guilty, and rightly so, if any of it be lost through negligence (13, 3).

By the Word of God Jesus became "flesh and blood" in the Incarnation. Likewise, by the Word of God—Jesus—bread and wine become "flesh and blood."

It is ironic that those who say they accept the Bible literally do not do so in the discourses on the Eucharist! Why? Because they know, consciously or subconsciously, that without apostolic succession they have no power to do what Christ wanted them to do!

But why do we still call the Body and Blood of Christ "bread" and "wine"? The answer is simple: After the consecration the appearances or accidents of bread and wine remain, but the reality, the substance, is the sacramental Body, Blood, Soul, and Divinity of Christ.

Let us finish with the words of the founder of Protestantism, Martin Luther (1517 AD). Even the founder of Protestantism had to admit to the historical truths of the Catholic Church's belief:

> *[Of] all the fathers, as many as you can name,* **not one** *has ever spoken about the sacrament as these fanatic [Protestants] do. None of the [early Christian writers] use such an expression as, 'It is simply bread and wine,' or 'Christ's body and blood are not present.' Yet [the subject of the Eucharist] is so frequently discussed by [the early Christian writers], it is impossible that they should not at some time have let slip such an expression as 'It is simply bread,' or 'Not that the body of Christ is physically present,' or the like, since they are greatly concerned not to mislead the people; actually, they simply proceed to speak as if no one doubted that Christ's body and blood are present. Certainly among so many fathers and so many writings a negative argument should have turned up at least once, as happens in other articles [of the faith]; but actually they all stand uniformly and consistently on the affirmative side (Luther's Works, St. Louis: Concordia Publishing, 1961, vol. 37, 54).*

Even Luther could not deny history!

It would have been wise for the Protestants to have taken the advice of the former Protestant, John Cardinal Henry Newman: "The Christianity of history is not Protestantism.... To be deep in history is to cease to be a Protestant."

Why do Catholics have a Mass?

Catholics have a Mass because Jesus instituted the Mass and the early Church always had a Mass. Let us look at an example from Luke's Gospel:

> *When the hour came, he took his place at table with the apostles. He said to them, "I have eagerly desired to eat this Passover with you before I suffer, for, I tell you, I shall not eat it [again] until there is fulfillment in the kingdom of God." Then he took a cup, gave thanks and said, "Take this and share it among yourselves; for I tell you [that] from this time on I shall not drink of the fruit of the vine until the kingdom of God comes." Then he took the bread, said the blessing, broke it, and gave it to them, saying, "This is my body,*

which will be given for you; do this in memory of me." And likewise the cup is the new covenant in my blood, which will be shed for you (Lk. 22:14-20).

When we look at history, the Mass is a well established reality for Christians. At first Christians celebrated Mass in their homes and with time they moved into public worship spaces, but the fundamental structure always remained the same.

It is astonishing to see in the year 150 AD, just 50 years after the death of the last apostle John, the existence in the Church of a set Mass structure that had to have been in place from the time of the apostles.

Justin Martyr, known by the friends of the apostles, wrote to the emperor Antononinus Pius in 150 about the long-standing practice of Christian worship in order to calm the anger and fear of the emperor in regard to the practices of the Christians.

Let us look at his description of the Mass in his letter:

On the day we call the day of the sun, all who dwell in the city or country gather in the same place, for it is on this day that the Savior Jesus Christ rose from the dead [In the early Church, according to Pliny, the Roman Governor of Pontus, in his Letters to the Emperor Trajan (ca. 111-113 AD,) the Christian faithful would often sing a "hymn to Christ as God" as they began their celebration of the "Lord's Supper."]

The memoirs of the apostles and the writings of the prophets are read, as much as time permits.

When the reader has finished, he who presides over those gathered admonishes and challenges them to imitate these beautiful things.

Then we all rise together and offer prayers* for ourselves...and for all others, wherever they may be, so that we may be found righteous by our life and actions, and faithful to the commandments, so as to obtain eternal salvation.

When the prayers are concluded we exchange the kiss.

The faithful, if they wish, may make a contribution and they themselves decide the amount. The collection is placed in the custody of the one who presides over the celebration to be used for the orphans, widows, and for any who are in need or distress.

Then someone brings bread and a cup of water and wine mixed together to him who presides over the brethren.

He takes them and offers praise and glory to the Father of the universe, through the name of the Son and of the Holy Spirit and for a considerable time he gives thanks (in Greek: eucharistian) that we have been judged worthy of these gifts.

When he has concluded the prayers and thanksgiving, all present give voice to an acclamation by saying: "Amen."
When he who presides has given thanks and the people have responded, those whom we call deacons give to those present the "eucharisted" bread, wine, and water and take them to those who are absent (Apol. 1, 65-67; PG 6, 428-429).

In explaining the mystery indicated by the word "eucharisted," Justin states in his *First Apology* (65):

We call this food Eucharist...since Jesus Christ our Savior was made incarnate by the word of God and had both flesh and blood for our salvation, so too, as we have been taught, the food which has been made into the Eucharist by the Eucharistic prayer set down by Him, and by the change of which our blood and flesh is nourished, is both the flesh and the blood of that incarnated Jesus.

Justin further goes on to say:

None is allowed to share in the Eucharist unless he believes the things which we teach are true...for we do not receive the Eucharist as ordinary bread and ordinary wine, but as Jesus Christ our Savior.

It is a wonder to me how anyone could be anything but Catholic! What you would have experienced in the year 150 and earlier is exactly what you experience today in any Catholic Church!

Is the Mass a true sacrifice?

*For I received from the Lord what I also handed on to you, that the Lord Jesus, on the night he was handed over, took bread, and, after he had given thanks, broke it and said, "This is my body that is for you. Do this in remembrance of me." In the same way he took the cup, and after supper, said, "This cup is the new covenant in my blood. Do this, as often as you drink it, in remembrance of me." For as often as you eat this bread and drink this cup, **you proclaim the death of the Lord** until he comes (1 Cor. 11:23-26).*

The expressions "This is my body, this is my blood" are taken from the Jewish language and theology of Temple sacrifice. For Jesus, these expressions designate himself as the *true and ultimate sacrifice*.

In the Old Testament, the Hebrew Scriptures, sacrifices of lambs, bulls, goats, and other animals were offered in the temple for the forgiveness of sins. Today, this sacrifice takes place in the mystery of the Mass, the bloodless sacrifice of the Lamb of God, Jesus Christ, at the altar of every Church, the New Temple of God. (It is no coincidence that John's Gospel has Jesus die at the exact time that the Jewish Temple sacrifices were taking place. It is Jesus who is the true Lamb, the true sacrifice. Jesus is the true Lamb that takes away the sins of the people).

The bloodless sacrifice of the Mass has traditionally been seen to have been prefigured in Genesis 14:18; 22:13, foretold in Malachi 1:10f, and attested to in 1 Corinthians 10:16, 18-21; 11:23-26 and Hebrews 13:10.

When the Jews were preparing for the Passover into the Promised Land, they offered up a paschal lamb and afterwards consumed the lamb, the victim, for strength for the journey (Ex. 12:1-20). This prefigures the Eucharistic sacrifice where Jesus, the Lamb of God, was offered up for our sins and then eaten sacramentally so that we may have the spiritual nourishment necessary to enter into the Promised Land of Heaven.

The Mass is a *re-presenting*, or making present of what took place once and for all at Calvary (Heb. 7:27; 9:12, 25-28; 10:10-14). Just as the Passover meal made present to those who participated in it the Exodus events, the Mass in a fuller way makes present what happened at Calvary. As Gregory of Nyssa (ca. 383) in his Sermon on the Resurrection (4) explains:

> *Jesus offered himself for us, Victim and Sacrifice, and Priest as well, and 'Lamb of God, who takes away the sin of the world.' When did he do this? When he made his own Body food and his own Blood drink for his disciples; for this much is clear enough to anyone, that a sheep cannot be eaten by a man unless its being eaten be preceded by its being slaughtered. This giving of his own Body to his disciples for eating clearly indicates that the sacrifice of the Lamb has now been completed.*

At every Mass Calvary is made present to us. Mass is a participation in that one and only sacrifice of Jesus on the cross at Calvary (cf. Heb. 7:27).

> *Our sin will not be small if we eject from the episcopate those who blamelessly...offered its Sacrifices (Clement, 4, trans. Jurgens). [W]e ought to do in order all things which the Master commanded us to perform at appointed times. He commanded us to celebrate sacrifices and services....at fixed times and hours (Ibid., 40, trans. Lake).*
>
> Clement of Rome (ca. 80)

The Council of Trent would affirm, in opposition to the Protestant Reformation, the belief of the early Church regarding the sacrificial nature of the Mass.

[Christ] our Lord and God, was once and for all to offer himself to God the Father by his death on the altar of the cross, to accomplish for them an everlasting redemption. But, because his priesthood was not to end with his death (cf. Heb. 7:24, 27), at the Last Supper, "on the night when he was betrayed" (1 Cor. 11:23), in order to leave to his beloved Spouse the Church a visible sacrifice (as the nature of man demands) "by which the bloody sacrifice which he was once and for all to accomplish on the cross would be re-presented, its memory perpetuated until the end of the world and its salutary power applied for the forgiveness of the sin which we daily commit"; declaring himself constituted "a priest for ever after the order of Melchizedek" (Ps. 110(109)4), He offered his body and blood under the species of bread and wine to God the Father, and, under the same signs...gave them to partake of to the disciples (whom he then established as priests of the New Covenant), and ordered them and their successors in the priesthood to offer, saying: "Do this as a memorial of Me," etc., (Lk. 22:19; 1 Cor. 11:24), as the Catholic Church has always understood and taught.

[After Christ] celebrated the old Pasch, which the multitude of the children of Israel offered...to celebrate the memory of the departure from Egypt (cf. Ex. 12:1f), Christ instituted a new Pasch, namely himself to be offered by the Church through her priests under visible signs in order to celebrate the memory of his passage from this world to the Father when by the shedding of his blood he redeemed us, "delivered us from the dominion of darkness and transferred us to His Kingdom" (cf. Col. 1:13).

This is the clean oblation which cannot be defiled by any unworthiness or malice on the part of those who offer it, and which the Lord foretold through Malachi would be offered in all places as a clean oblation to his name (cf. Mal. 1:11). The apostle Paul also refers clearly to it when, writing to the Corinthians, he says that those who have been defiled by partaking of the table of devils cannot be partakers of the table of the Lord. By "table" he understands "altar" in both cases (cf. 1 Cor. 10:21). Finally, this is the oblation which was prefigured by various types of sacrifices under the regime of nature and of the law (cf. Gen. 4:4; 8:20; 12:8; Ex. passim). For it includes all the good that was signified by those former sacrifices; it is their fulfillment and perfection.... (J. Neuner and J. Dupuis, eds. The Christian Faith: Doctrinal Documents of the Catholic Church, New York: Alba House, 1990), ND 1546-1547).

It is no coincidence that the Eucharist as the Body and Blood of Christ as well as Jesus' sacrifice and death on the cross were considered foolishness to many and a stumbling block to others (cf. Jn. 6:60; 1 Cor. 1:23). To believe in the Eucharist and in the death and resurrection of Jesus can only occur through a gift from God (cf. Jn. 6:65).

- The Greek for "Do this in memory of me" (Luke 22:19), *Touto poieite tan eman anamnasin*, can also be translated as "Offer this as a memorial offering." The *Didache* often applies the Greek word *thusia*, or sacrifice when referring to the Eucharist.

Why do we celebrate the Lord's Day on a Sunday?

The Seventh Day Adventists are the ones who most often condemn Catholics and Protestants of various denominations for celebrating the Sabbath on Sunday. They argue that this is a violation of Exodus 20:8-11 where the Sabbath is designated as Saturday. They also point out that Christ went to worship on the Sabbath on Saturday (Lk. 4:16; Lk. 23:56). Hence, if Christ worshiped on the Sabbath, on Saturday, we are called to do the same.

As Catholics we respond by reminding them that the Sabbath, the "Lord's Day" was eventually moved in the early Church by the apostle Peter to Sunday because Jesus Christ rose on a Sunday (Rev. 1:10; Acts 20:7).

In the *Didache*, also known as *The Teachings of the Twelve Apostles*, (9) we read:

On Sunday, the Lord's day, break bread and give your Eucharistic thanks....

The Bible also commands the obligatory sacrifice of animals (Gn. 4:4; Lv. 1:14), the following of dietary Kosher practices (cf. Deut. 12:15-28; 14:3-21) and circumcision (Gn. 17:10; Lk. 2:21). Jesus was circumcised and he offered sacrifices in the Temple. Yet I don't see Seventh Day Adventists sacrificing animals, or for that matter, following all the Levitical laws that Jesus would have observed!

Jesus is the "Lord of the Sabbath" (Mt. 12:8) and he entrusted his authority over the Sabbath and other things to his Church, his Body (cf. Mt. 10:40; 16:19; 18:18-20; Lk. 10:16). Furthermore, Jesus reminds us that the Spirit will guide us on how and when to pray (cf. Rom. 8:26-27).

Just as Peter was empowered to change the dietary laws (cf. Acts 10:9-33), and just as Peter, James and the rest of the apostles were empowered to eliminate the demand for circumcision for converts (cf. Acts 15:1-35) the Church, the Body and Bride of Christ (1 Cor. 12:12f; 2 Cor. 11:2; Rom. 12:5; Eph. 1:22f; 5:25; Rev. 19:7), was empowered to change the day for the Sabbath.

The early Church recognized that the true Sabbath was now to be celebrated in Christ and it was to be held on Sunday, the Lord's Resurrection

Day. Sunday came to be understood as the *first* day and the *eighth* day, the eighth day signifying perfection.

The Mass, completely Biblical

> *In your Eucharistic assemblies, in the holy churches [dioceses], after all good patterns form your gatherings, arrange the places for the brethren carefully with all sobriety. Let a place be reserved for the presbyters [priests] in the midst of the eastern part of the house, and let the throne of the bishop be placed amongst them; let the presbyters sit with him; but also at the other eastern side of the house let the laymen sit; for thus it is required that the presbyters should sit at the eastern side of the house with the bishops, and afterwards the laymen, and next the women; that when you stand to pray the rulers may stand first, afterwards the laymen, and then the women also, for towards the East it is required that you should pray, as you know that it is written, 'Give praise to God who rideth on the heavens of heavens towards the East (Ps. 68). As for the deacons, let one of them stand constantly over the gifts of the Eucharist, and let another stand outside the door and look at those who come in; and afterwards when you make offerings, let them serve together in the Church. And if a man be found sitting out of his place, let the deacon who is within reprove him, and make him get up and sit in the place that befits him (12).*
>
> *Didascalia Apostolorum (ca. 50-251?)*

Most Catholics and non-Catholics are unaware of the Biblical foundation for the prayers within the Mass. They may understand that some of the prayers are biblically oriented, but are often surprised to find out that every prayer within the Mass, every part of the Mass, every word within the Mass is an expression of the very words of Scripture. The following will illustrate the point:

The Introductory Rites of the Mass
Priest: In the name of the Father, and of the Son, and of the Holy Spirit (cf. Mt. 28:19).
Congregation: Amen (cf. 1 Chr. 16:36).

Greeting by Priest
[Form A] The grace of our Lord Jesus Christ and the Love of God and the fellowship of the Holy Spirit be with you all (cf. 2 Cor. 13:13).

[Form B] The grace and peace of God our Father and the Lord Jesus Christ be with you (cf. 1 Pet. 1:3).

[Form C] The Lord be with you (cf. Ruth 2:4).

Penitential Rite

Priest and Congregation: I confess to almighty God, and to you, my brothers and sisters, that I have sinned through my own fault (cf. James 5:16) in my thoughts and in my words (Rom. 12:16; James 3:6) in what I have done, and in what I have failed to do; (James 4:17) and I ask Blessed Mary, ever virgin, all the angels and saints, and you, my brothers and sisters, to pray for me to the Lord our God (cf. 1 Thess. 5:25).

Priest: May almighty God have mercy on us, forgive our sins, and bring us to everlasting life (cf. 1 Jn. 1:9).

Priest and Congregation: Lord have mercy. Christ have mercy. Lord have mercy (cf. Tob. 8:4; 1 Tim. 1:2).

Gloria

Priest and Congregation: Glory to God in the highest, and peace to his people on earth (cf. Lk. 2:14). Lord God, heavenly King, almighty God and Father, (Rev. 19:6) we worship you (Rev. 22:9), we give you thanks (Eph. 5:20), we praise you for your glory (Rev. 7:12). Lord Jesus Christ, only Son of the Father (2 Jn. 3), Lord God, Lamb of God, you take away the sin of the world: have mercy on us (cf. Jn. 1:29); you are seated at the right hand of the Father: receive our prayer (Rom. 8:34). For you alone are the Holy One (cf. Lk. 4:34), you alone are the Lord (Rev. 15:4), you alone are the most High, Jesus Christ (Lk. 1:32), with the Holy Spirit, in the glory of God the Father. Amen (cf. Jn. 14:26).

Profession of Faith

Priest and Congregation: We believe in God, the Father, Almighty, maker of heaven and earth (cf. Gen. 14:19), of all that is seen and unseen (Col. 1:16). We believe in one Lord, Jesus Christ, the only Son of God, eternally begotten of the Father (Lk. 1:35), God from God, Light from Light, true God from true God, begotten, not made, one in Being with the Father (Heb. 1:3). Through him all things were made (Jn. 1:1-4). For us men and for our salvation he came down from heaven (cf. Jn. 3:13): by the power of the Holy Spirit he was born of the Virgin Mary, and became man (cf. Mt. 1:18). For our sake he was crucified under Pontius Pilate (Jn. 19:16); he suffered, died, and was buried. On the third day he rose again in fulfillment of the Scriptures (cf. 1 Cor. 15:3); he ascended into heaven (cf. Lk. 24:51) and is seated at the right hand of the Father (cf. Col. 3:1). He will come again in glory to judge the living and the dead (cf. 2 Tim. 4:1), and his kingdom will have no end (Lk. 1:33). We believe in the Holy Spirit, the Lord, the giver of Life (Acts 2:17), who proceeds from the Father and the Son. With the Father and the Son he is worshiped and glorified (Jn. 14:16). He has spoken through the Prophets (1 Pet. 1:10-11). We believe in one holy catholic and apostolic Church (cf. Rom. 12:5). We acknowledge one baptism for the forgiveness of sins (Acts 2:38). We look for the resurrection of the dead, and the life of the world to come. Amen (cf. Rom. 6:5).

Liturgy of the Eucharist

Priest: Blessed are you, Lord, God of all creation. Through your goodness we have this bread to offer, which earth has given and human hands have made (cf. Qo 3:13). It will become for us the bread of life (cf. Jn. 6:35). Blessed are you, Lord, God of all creation. Through your goodness we have this wine to offer, fruit of the vine and work of human hands. It will become our spiritual drink (cf. Lk. 22:17-18).

Congregation: Blessed be God forever (cf. Ps. 68:36).

Priest: Pray, brethren, that our sacrifice may be acceptable to God, the almighty Father (cf. Heb. 12:28).

Congregation: May the Lord accept this sacrifice at your hands for the praise and glory of his name, for our good and the good of all his Church (cf. Ps. 50:23).

Eucharistic Prayer

Priest: Lift up your hearts.
Congregation: We lift them up to the Lord (Lam. 3:41).
Priest: Let us give thanks to the Lord our God (Col. 3:17).
Congregation: It is right to give him thanks and praise (Col. 1:3).

Preface Acclamation

Priest and Congregation: Holy, holy, holy, Lord, God of power and might, heaven and earth are full of your glory. Hosanna in the highest. Blessed is he who comes in the name of the Lord. Hosanna in the highest (cf. Is. 6:3; Mk. 12:9-10).

Eucharistic Prayer I

We come to you, Father, with praise and thanksgiving, through Jesus Christ your Son (cf. Eph. 5:20).

Through him we ask you to accept and bless these gifts we offer you in sacrifice (cf. 2 Macc. 1:26).

We offer them for your holy Catholic Church, watch over it, Lord, and guide it; grant it peace and unity throughout the world. We offer them for N. our pope, for N. our bishop, and for all who hold and teach the Catholic faith that comes to us from the apostles (cf. Jn. 17:21; Acts 2:42).

Remember, Lord, your people, especially those for whom we now pray, N. and N. Remember all of us gathered here before you. You know how firmly we believe in you and dedicate ourselves to you. We offer you this sacrifice of praise for ourselves and those who are dear to us. We pray to you, our living and true God, for our well-being and redemption (cf. Ps. 106:4; Heb. 13:15).

In union with the whole Church, we honor Mary, the ever-virgin mother of Jesus Christ our Lord and God. We honor Joseph, her husband, the apostles and martyrs, Peter and Paul, Andrew…and all the saints. May their merits and prayers gain us your constant help and protection (cf. Mt. 1:2-16; Lk. 16:9; 1 Cor. 12:12, 20f; Rev. 5:8).

Father, accept this offering from your whole family. Grant us your peace in this life, save us from final damnation, and count us among those you have chosen (cf. Col. 1:11).

Bless and approve our offering; make it acceptable to you, an offering in spirit and in truth (Jn. 4:24).

Let it become for us the body and blood of Jesus Christ, your only Son, our Lord.

The day before he suffered he took bread in his sacred hands and looking up to heaven, to you, his almighty Father, he gave you thanks and praise. He broke the bread, gave it to his disciples, and said:

Take this, all of you, and eat it: this is my body which will be given up for you (cf. Mt. 26:26-28).

When supper was ended, he took the cup. Again he gave you thanks and praise, gave the cup to his disciples, and said:

Take this, all of you, and drink from it: this is the cup of my blood, the blood of the new and everlasting covenant. It will be shed for you and for all so that sins may be forgiven.

Do this in memory of me (Mt. 26:26-28).

Let us proclaim the mystery of faith (cf. 1 Tim. 3:16):
Christ has died, Christ is risen, Christ will come again (cf. 1 Cor. 15:3-5).

Father, we celebrate the memory of Christ, your Son. We, your people and your ministers, recall his passion, his resurrection from the dead, and his ascension into glory; and from the many gifts you have given us we offer to you, God of glory and majesty, (cf. 1 Pet. 1:18-21) this holy and perfect sacrifice (Heb. 9:13-14): the bread of life and the cup of eternal salvation (Jn. 6:54).

Look with favor on these offerings and accept them as once you accepted the gifts of your servant Abel (Gen. 4:4), the sacrifice of Abraham, our father in faith (Gen. 22:12), and the bread and wine offered by your priest Melchisedech (Gen. 14:18).

Almighty God, we pray that your angel may take this sacrifice to your altar in heaven (cf. Rev. 8:3-4). Then, as we receive from this altar the sacred body and blood of your Son, let us be filled with every grace and blessing (cf. Eph. 1:3).

Remember, Lord, those who have died and have gone before us marked with the sign of faith, especially those for whom we now pray, N. and N. May these, and all who sleep in Christ, find in your presence light, happiness, and peace (cf. 1 Thess. 4:13, 14).

For ourselves, too, we ask some share in the fellowship of your apostles and martyrs, with John the Baptist, Stephen, Matthias, Barnabas...Felicity, Perpetua, Agatha... and all the saints (cf. Col. 1:12).

Though we are sinners, we trust in your mercy and love. Do not consider what we truly deserve, but grant us your forgiveness (cf. Ps. 25:7).

Through him you give us all these gifts. You fill them with life and goodness, you bless them and make them holy (Ps. 104:27-28).

*The following Eucharistic prayers find their expressions in the following Scripture passages:
Eucharistic Prayer II (cf. 2 Macc. 14:36; Phil. 2:8; Jn. 10:17-18; Mk. 14:22-25; Heb. 2:14-15; Jn. 6:51; 1 Cor. 10:17; 2 Macc. 12:45-46; 2 Thess. 1:4-5)

Eucharistic Prayer III (cf. Tob. 8:5; Jn. 1:3; Ps. 113:3; Lk. 22:19-20; 1 Cor. 11:26; Mk. 13:33; 2 Cor. 5:19; Eph. 4:3; Eph. 5: 25-27; Jn. 17:22, 23; Col. 1:4-5)

Eucharistic Prayer IV (cf. Gen. 1:26; Is. 55:6; 55:3; Gal. 4:4-5; Heb. 4:15; Lk. 4:18; 1 Cor. 15:54-57; Jn. 14:16; Heb. 9:15; Jn. 13:1; 1 Cor. 11:23-25; Jn. 4:42; 1 Pet. 3:18, 19; Eph. 1:19-20; Mt. 25:31; 1 Cor. 12:12, 27; Acts 10:35; Rom. 8:20-21)

Doxology
Priest: Through him, with him, in him, in the unity of the Holy Spirit, all glory and honor is yours, almighty Father, for ever and ever. Amen (cf. Rom. 11:36).

Communion Rite
Priest and Congregation: "Our Father" (cf. Mt. 6:9-13).
Priest: Deliver us, Lord, from every evil, and grant us peace in our day. In your mercy keep us free from sin and protect us from all anxiety as we wait in joyful hope for the coming of our Savior Jesus Christ (Jn. 17:15).
Congregation: For the kingdom, the power, and the glory are yours, now and for ever.
Priest: Lord Jesus Christ, you said to your apostles: I leave you peace, my peace I give you.

Look not on our sins, but on the faith of your Church, and grant us the peace and unity of your kingdom where you live and reign for ever and ever (Jn. 14:27). The peace of the Lord be with you always (cf. Jn. 20:19).

Breaking of the Host
Priest and Congregation: Lamb of God, you take away the sins of the world, have mercy on us.... Lamb of God, you take away the sins of the world, grant us peace (cf. Jn. 1:29).

Communion
Priest: This is the Lamb of God who takes away the sins of the world. Happy are those who are called to his supper (Rev. 19:9).
Priest and Congregation: Lord, I am not worthy to receive you, but only say the word and I shall be healed (cf. Mt. 8:8).

Dismissal
Priest: The Lord be with you.
Congregation: And also with you (cf. Ruth 2:4).
Priest: May almighty God bless you, the Father, and the Son, and the Holy Spirit (cf. Lk. 24:51). Go in the peace of Christ (cf. Lk. 7:50; 2 Chron. 35:3).
Congregation: Thanks be to God (cf. 2 Cor. 9:15).

(For a more detailed analysis see Peter M. J. Stravinskas, *The Catholic Church and the Bible* (San Francisco: Ignatius Press, 1987): 83-106).

Where do Catholics get the idea of mortal and venial sins?

Aren't all sins just the same? Sin is sin! Where do Catholics get this idea of mortal and venial sin? Where do they get this idea that some sins are more serious than others?

Well, common sense tells us that some sins are more serious than others. Murder is much more serious than stealing bubble gum, isn't it?

Now let us look to the Scriptures. In 1 John 5:16-17 we see that John makes reference to some sins as being "deadly" and some sins as being "not deadly." Some sins are deadly, mortal, and some sins are venial, not deadly. A mortal, or deadly sin, is one that turns a person away from God, and a venial, not deadly, sin is one that wounds, but does not destroy one's relationship with God.

Other examples of losing grace by committing serious sins are found in Romans 11:21-22, Hebrews 10:26-31; and 2 Peter 2:20-22. We are reminded that if we do not remain in his kindness, his grace, we "will be cut off" (Rom. 11:21-22). If we sin deliberately and seriously we can expect the fearful prospect of judgment and the flaming and consuming fire of damnation (cf. Heb. 10:26f). If we fall to the defilements of the world, then our last condition will be damnable (cf. 2 Pet. 2:20f).

Why do we need priests to forgive serious sins?

Father, I have sinned against heaven and against you (Lk. 15:21).

You are to confess your sins in the Church. This is the way of life.
Didache 4, 12, 14 (ca. 65)

Serious sin, or what we call mortal/deadly sin (1 Jn. 5:17) requires the authority of the priest as an authoritative, power-filled representative of God and of the community. When we look at the Scriptures (Mt. 18:18; 16:19; Jn. 20:21-23) it becomes obvious that God entrusted his apostles with the gift of forgiving sins. In the words of the apostle John:

Jesus said to [the apostles], "Peace be with you. As the Father has sent me, even so I send you." And when he had said this, he breathed on them, and said to them, "Receive the Holy Spirit. If you forgive the sins of any, they are forgiven; if you retain the sins of any, they are retained" (Jn. 20:21-23, RSV).

Notice that Jesus didn't say "Now go out into the world and tell people to confess their sins directly to God and he will forgive everyone's sins." Rather he said, "If **you** forgive the sins of any, they are forgiven; if **you** retain the sins of any, they are retained." Jesus empowered the apostles to forgive sins in his name! (Notice Jesus "breathed" on the apostles. Throughout the Scriptures the breath of God is associated with new life and God's creative work. The breathing on the apostles made them priests, the first bishops, and made them able to forgive sins *in persona Christi capitas*. And by forgiving sins they were making those forgiven into "new creations" in Christ.)

Pacian of Barcelona (392 AD) notes the necessity of priests for the forgiveness of sins in his *Sermon on Penance*:

Certainly God never threatens the repentant; rather, he pardons the penitent. You will say that it is God alone who can do this. True enough; but it is likewise true that he does it through his priests, who exercise his power. What else can it mean when he says to his apostles: 'Whatever you shall bind on earth shall be bound in heaven; and whatever you shall loose on earth shall be loosed in heaven?' Why should he say this if he were not permitting men to bind and loose? And he clearly was not permitting this to the apostles alone? Were that the case, he would likewise be permitting them alone to baptize, them alone to confer the Holy Spirit in confirmation, them alone to cleanse the pagans of their sins; for all of these things are commissioned not to others but to the apostles. But if the loosing of bonds and the power of the Sacrament is given to anyone in that place, either the whole is passed on to us priests from the form and power of the apostles, or nothing of it can be imparted to us priests by whatever decrees. If, then, the power both of baptism and confirmation, greater by far than charisms, is passed on to bishops and priests by apostolic succession, so too is the right of binding and loosing (1, 6).

Jesus has an important reason for giving us the Sacrament of Penance. When we sin we harm our relationship with God, the community, and we do damage to ourselves (cf. Lk. 15:21). That is because when we sin we break the commandments that God placed side by side, the love of God and the love of neighbor as ourselves (Mt. 22:37-40). For example, if one steals one dulls one's conscience, hurts the person whose property was stolen, and breaks God's seventh commandment.

Since sin damages our relationship with God, our relationship with ourselves, and our relationship with others, it needs to be healed in all three dimensions.

The priest—as a member of the human race—therefore is a representative of God and of the community and he brings Christ's healing and the community's healing, as the Body of Christ, to the sinner. That is why God chose the apostles, the first bishops, the first priests, to forgive sins.

When Jesus said to Peter, "Whoever's sins you bind shall be bound, and whoever's sins you loose shall be loosed" (cf. Mt. 16:18f), he was saying— within the Judaic and Hebrew understanding of the terms "bound" and "loose"—whoever you exclude from your communion will be excluded from communion with God and whoever you receive into your communion God will welcome back into his. Reconciliation with God is inseparable from reconciliation with the Church (cf. 1 Cor. 12:12f; Rom. 12:5; Eph. 1:22f; 1 Cor. 3:9, 10, 16; 1 Thess. 1:4; 1 Tim. 3:5, 15) (CCC 1445).

Christ forgives sins by means of priests in the Sacrament of Penance because God gave that authority to the apostles and their successors (Mt. 18:18; 16:19; Jn. 20:21-23). Paul reminds the faithful that he has been entrusted with the "ministry of reconciliation" (2 Cor. 5:18-20); James reminds us, within the context of the Sacrament of Anointing, that the presbyter, the priest, administers Christ's forgiveness (Jms. 5:14-16).

God knew that a human person acting in the Person of Christ, *in persona Christi capitas*, or as Another Christ, *alter Christus* (cf. Mt. 10:40; Lk. 10:16; Lk. 25:47), could bring the only true healing that people need to have. People need to hear from someone they are forgiven. I don't know how many times a person has broken down crying after having his or her sins forgiven after a priest has given the gift of Christ's absolution. The sense of being created anew is miraculous for that person.

Studies in the past, when Catholics practiced going to Confession on a weekly or monthly basis, noted that Catholics had the lowest rate of psychological disorders in America. It is interesting to cite that in this modern era where Catholics have abandoned the frequent use of the

Sacrament of Penance the rate of psychological disorders by American Catholics has increased to match that of the rest of the American population.

Ignatius of Antioch (ca. 107), the disciple of the apostle John, recognized the importance and absolute necessity of confession to a priest when he said: "The Lord...forgives all who repent, if their repentance leads to the unity of God and the council of the bishop" (*Philadelphia*, 8, trans. Lake). In Cyprian of Carthage's *Letter to the Clergy* (ca. 250) [cf. 16 (9), 2] he writes: "Sinners may come to confession and, through the imposition of hands by the bishop and priests, may receive re-admittance into the life of the Church." And in his letter to *The Lapsed* (ca. 351) (28) Cyprian writes: "I beseech you, brethren, let everyone who has sinned confess his sin while he is still in this world, while his confession is still admissible, while satisfaction and remission made through the priests are pleasing before the Lord."

The Sacrament of Penance is a healing sacrament. Why do we confess to priests our mortal sins? Because Christ commanded it!

The spiritual effects of the Sacrament of Penance are beautifully summarized in the *Catechism of the Catholic Church* (1496):

> -reconciliation with God by which the penitent recovers grace;
> -reconciliation with the Church;
> -remission of the eternal punishment incurred by mortal sins;
> -remission, at least in part, of temporal punishments resulting from sin;
> -peace and serenity of conscience, and spiritual consolation;
> -an increase of spiritual strength for the Christian battle.

Theodore of Mopsuestia (ca. 428) reminds us of the above. He reminds us that the priest is a father (as in 1 Cor. 4:14-15; 1 Tim. 1:2; Tit. 1:4; Philem. 10; 1 Thess. 2:1) who takes care of his children, a spiritual doctor that brings healing to souls:

> If we commit a great sin against the commandments we must first induce our conscience with all our power to make haste and repent of our sins as is proper, and not permit ourselves any other medicine. This is the medicine for sins, established by God and delivered to the priests of the Church, who make diligent use of it in healing the affliction of men. You are aware of these things, as also of the fact that God, because he greatly cares for us, gave us penitence and showed us the medicine of repentance; and he established some men, those who are priests, as physicians of sins. If in this world we receive through them healing and forgiveness of sins, we shall be delivered from the judgment that is to come. It behooves us, therefore, to draw near to the priests in great confidence and to reveal to them our sins; and those priests, with all diligence,

solicitude, and love, and in accord with the regulations mentioned above, will grant healing to sinners. The priests will not disclose the things that ought not to be disclosed; rather, they will be silent about the things that have happened, as befits true and loving fathers who are bound to guard the shame of their children while striving to heal their bodies... (Catechetical Homilies, 16).

The Church has always, from the beginning of the Church, had confession of sins to priests.

- The priest acts in the Person of Christ the Head, in the second person of the Trinity, in the place of God, for only God can forgive sins (cf. Mk. 2:7). It is Christ, God, forgiving sins through the priest.

What does it mean to be excommunicated?

The Catholic Catechism describes excommunication in the following manner in section 1463:

Certain particularly grave sins incur excommunication, the most severe ecclesiastical penalty, which impedes the reception of the sacraments and the exercise of certain ecclesiastical acts, and for which absolution consequently cannot be granted, according to canon law [canon 1331], except by the pope, the bishop of the place or priests authorized from them.

One can be excommunicated by means of an ecclesiastical trial or proceeding. One can also—due to the extreme severity of an offense—be excommunicated automatically. The procurement of an abortion is such an example of where an automatic excommunication takes place at the moment of the action.

From first appearances, the idea of being excommunicated may seem quite frightening, and it should. But the real purpose of excommunicating an individual or an individual excommunicating himself or herself is to call that person to repentance. Excommunication is a call to come back home into the fold of Christ and his Body, the Church. Excommunication is the Church's way of warning people about their eternal destiny. It is a way of warning people of the consequences of their actions.

When one recognizes one's wrongful act or acts, and repents, one may have his or her sentence of excommunication lifted by a pope, a bishop, or a priest with the proper authorization.

Examples of excommunication in the Scriptures can be found in 1 Corinthians 5:3-5, 9-13, in 2 Thessalonians 3:6,14, in 1 Timothy 1:20 and in

Titus 3:10f. Protecting the souls of the faithful is at the heart of excommunication. Paul reminds the Corinthians to "purge the evil from [their] midst" (5:13) and to "deliver [the unrepentant evil] to Satan for the destruction of [their] flesh" (5:5). In his second letter to the Thessalonians the community is reminded "shun any brother who conducts himself in a disorderly way and not according to the tradition they received from us" (3:6), and "if anyone does not obey our word...take note of this person so as not to associate with him" (3:14). In 1 Timothy we read how Hymenaeus and Alexander have been "handed over to Satan to be taught not to blaspheme" (20). And in Titus 3:10f we read: "After a first and second warning, break off contact with a heretic, realizing that such a person is perverted and sinful and stands self-condemned" (11).

Excommunication is the Church's way of warning people about their eternal destiny. It is a way of warning people of the consequences of their actions.

Why indulgences?

For he is purged as if by certain works of the whole people, and is washed in the tears of the multitude; by the prayers and tears of the multitude he is redeemed from sin, and is cleansed in the inner man. For Christ granted to His Church that one should be redeemed through all, just as His Church was found worthy of the coming of the Lord Jesus so that all might be redeemed through one (1, 15, 80).
Ambrose of Milan (ca. 333)

The *Catholic Encyclopedia* describes indulgences in the following manner:

Remission of the temporal punishments for sins, and therefore the giving of satisfaction owed God for one's sin. Indulgences are granted either after the sacrament of Penance or by perfect contrition. Indulgences are either plenary (when all punishments are remitted) or partial (when only part of that punishment is remitted). Plenary indulgences demand that one be free of all venial sin, but partial indulgences do not require this.

Partial indulgences remit that amount of temporal punishment that would be remitted in the ancient Church by performances for the designated period of time. Indulgences can only be gained for oneself or for those in purgatory, but not for other living human beings. Indulgences are derived from the treasure of merits of the saints, from Christ Himself or from His Mother (Catholic

Encyclopedia, ed. Peter M. J. Stravinskas, Huntington: Our Sunday Visitor, Inc., 1991, 509).

As members of the Church we make up the Body of Christ, with Christ as the Head. Because of this reality we share in the life of the whole Church, the Church on earth, in purgatory, and in heaven. As one Body we profit from the prayer and good works of others. As Paul mentions in Romans 12:4-8:

> *Just as each of us has various parts in one body, and the parts do not all have the same function: in the same way, all of us, though there are so many of, make up one body in Christ, and as different parts we are all joined to one another. Then since the gifts that we have differ according to the grace that was given to each of us: if it is a gift of prophecy, we should prophesy as much as our faith tells us; if it is a gift of practical service, let us devote ourselves to serving; if it is teaching, to teaching; if it is encouraging, to encouraging (NJB).*

This sense of interconnectedness in prayer, works, and gifts is also described beautifully by Ambrose (ca. 340) in his *Treatise on Cain:*

> *You are told to pray especially for the people, that is, for the whole body, for all its members, the family of your mother the Church; the badge of membership in this body is love for each other. If you pray only for yourself, you pray for yourself alone. If each one prays for himself, he receives less from God's goodness than the one who prays on behalf of others. But as it is, because each prays for all, all are in fact praying for each one. To conclude, if you pray only for yourself, you will be praying, as we said, for yourself alone. But if you pray for all, all will pray for you, for you are included in all. In this way there is a great recompense; through the prayers of each individual, the intercession of the whole people is gained for each individual. There is here no pride, but an increase in humility and a richer harvest from prayer* (Cf. *Lib.* 1, 9, 34, 38-39: CSEL 32, 369. 371-372).

We are all indispensable to one another in our prayers and gifts. What one is lacking another has an extra amount of, and vice versa.

Now let us look at Colossians 1:24 where Paul states,

> *I rejoice in my sufferings for your sake, and in my flesh I am filling up what is lacking in the afflictions of Christ on behalf of his body, which is the church....*

Paul is suffering for the Body, the Church. This is an allusion to how the debt for sin can be made up for another.

And when we look to the above citations with reference to 1 John 2:2 (where Jesus expiates sins) we recognize, by the merits of Christ, the Church, his Body, his Bride (cf. Mk. 2:19; Lk. 5:34), we have an inexhaustible fund for the payment or satisfaction of sins.

Likewise, all the saints, and particularly the greatest saint of all, Mary, by their prayers and works and sufferings have built up a reservoir of prayers and works for others, a reservoir of satisfaction. Again as Paul reminds us: *"I rejoice in my sufferings for your sake, and in my flesh I am filling up what is lacking in the afflictions of Christ on behalf of his body, which is the church..."* (Colossians 1:24). Christians have been blessed by their savior in sharing in his redemptive work!

When we take these Scriptural principles together we come up with the Church's understanding of indulgences, and why the Church has always recognized the reality of indulgences.

Jesus expiates sins (1 Jn. 2:2) and therefore his body, the Church, which is inseparable from its head, also expiates sins (Rom. 12:4-8)—for one cannot decapitate the head from the body. When we put these two quotes together with Paul's insight in Colossians 1:24 that states that we make up in our sufferings "what is lacking in the afflictions of Christ on behalf of the body, which is his Church," then we can see how the Church can possess a spiritual reservoir of satisfaction for the good of others. For nothing spiritual is ever wasted; It always finds a home!

Christ forgives sins. When one's sins are forgiven the guilt is completely washed away, completely forgiven and forgotten. Yet divine justice demands that the injury that results from sin be repaired. If one has murdered a person and repents and seeks God's forgiveness, God shall bring that forgiveness, in this particular case, through the Sacrament of Penance.

While Christ thus forgives the person of all his or her guilt, the world is injured by the loss of a person whose contributions to the world have been lost. I think in particular of an account told by a woman regarding the confession of the sin of abortion to Padre Pio, the great Stigmatist and saint. Padre Pio conferred God's forgiveness upon her, but he reminded her that the world had lost in that aborted child a future pope. The sin is forgiven, but the damage from the sin lingers on.

It is this lingering damage that needs to be paid off. The greater the damage, the greater the payment. In the Sacrament of Penance the eternal punishment for grave sin is forgiven, but the temporal punishment still awaits payment—sometimes the penance imposed by the priest suffices at other times more is required.

In the case of very holy people, the penance expiates or pays off the temporal punishment, the lingering damage. In the case of less holy individuals, the penance pays part of the payment for the lingering damage. Again, the sin is forgiven, heaven is guaranteed, but if one has not paid the whole debt off in this life, one pays it off in purgatory.

This sense of temporal or lingering punishment that needs to be cleansed, even after God's forgiveness, is seen in 2 Samuel 12:14f. David is completely forgiven by God, but still pays a price for his sin. Nathan said, "The Lord on his part has forgiven your sin.... But since you have utterly spurned the Lord by this deed, the child born to you must surely die."

Let us recognize the giftedness of our prayers, works, and sufferings for the good of ourselves or those loved ones in purgatory. The next time we have a difficult day at work, or feel ill, let us offer it up for the good of our soul and/or the good of the souls in purgatory.

The problem with indulgences, prior to the Protestant Reformation, was not so much a problem dealing with the belief as much as a problem dealing with the abuse of the use of indulgences. Sinners often paid professional penitents to expiate their temporal punishments; some priests sold indulgences without pointing out the requirements that were associated with indulgences; and some priests sold indulgences to raise money to build churches. The right motives and the right understanding of the role of indulgences were not present in these abuses.

When Pius V refused to grant any indulgences that had to do with any form of monetary transactions, the abuses eventually disappeared.

Today indulgences have returned to the intended purpose, a gift from the "Treasury of the Church."

Is there a Sacrament of Holy Orders?

> *Let the bishop be ordained after he has been chosen. When someone pleasing to all has been named, let the people assemble on the Lord's Day with the presbyters and with such bishops as may be present. All giving assent, the bishops shall impose hands on him, and the presbytery shall stand in silence (2). When the presbyter is to be ordained, the bishop shall impose his hand upon his head while the presbyters touch the one to be ordained....(8). When a deacon is to be ordained the bishop alone shall lay his hands upon him (9).*
> *Hippolytus of Rome (ca. 200)*

The Sacrament of Holy Orders is an indispensable part of the Church. Without it the Church could not trace itself back to apostolic times, and therefore back to Christ.

As in Old Testament times (cf. Ex. 19:6; Num. 18:1-7), the Church makes a distinction between the common priesthood of all the faithful (1 Pet. 2:9) and the ordained priesthood. The Levitical priesthood would be replaced by Jesus by his own priesthood and his own priests. Through the providential mystery of God the ancient temple where sacrifices were performed by the Levitical priests was destroyed in 70 AD by the Romans, never to be rebuilt! Thus the new priests would be the priests of the New Covenant, the priests according to the order of Melchizedek, priests who act in the person of Jesus Christ himself.

All Christians are called to be a priestly people, a healing, loving, forgiving people, but some of the faithful were specifically set aside by Jesus and the apostles for unique ministerial roles.

The priesthood conferred by the Sacrament of Holy Orders is one that is specifically designated for teaching, leading worship, and meeting the pastoral needs of the people. Holy Orders confer an indelible spiritual mark on the soul.

The most important of the Holy Orders is that of the bishop because he serves as the visible head of the local or particular church (cf. 1 Tim. 3:1-7; Titus 1:7). Every bishop in the world can trace himself from one bishop to another bishop to another bishop all the way back in time to an apostle. Consequently, they have the fullness of the priesthood and are crucial in protecting the true faith. The greatest of the bishops is of course the pope, since he is the successor of the leader of the apostles, Peter.

The next order is the order of presbyter or what we commonly call the priest (cf. 1 Timothy 5:17f). He is a "prudent-coworker" and extension of the bishop. He receives his authority from the bishop, and teaches in power because of his tie to the tree of apostolic succession.

The final order is that of the deacon who likewise is attached to the bishop, but who is entrusted primarily with works of charity (cf. Acts 6:1-7; 1 Timothy 3:8-13).

Holy Orders were instituted by Christ (Lk. 22:19; Jn. 20:22f), conferred by the imposition of hands by an apostle or his successor (Acts 6:6; 13:3; 14:23), and give grace (1 Tim. 4:14; 2 Tim.1:6-7).

Clement of Rome, the friend of the apostle Peter, eloquently teaches us about the gift of the priesthood.

The apostles preached to us the Gospel received from Jesus Christ, and Jesus Christ was God's Ambassador. Christ, in other words, comes with a message from God, and the apostles with a message from Christ. Both these orderly arrangements, therefore, originate from the will of God. And so, after receiving their instructions and being fully assured through the Resurrection of our Lord Jesus Christ, as well as confirmed in faith by the word of God, they went forth, equipped with the fullness of the Holy Spirit, to preach the good news that the Kingdom of God was close at hand. From land to land, accordingly, and from city to city they preached, and from among their earliest converts appointed men whom they had tested by the Spirit to act as bishops and deacons for future believers. And this was no innovation, for, a long time before the Scriptures had spoken about bishops and deacons, for somewhere it says: I will establish overseers in observance of the law and their ministers in fidelity (Clement of Rome, *Epistle to the Corinthians,* 42, quoted in *The Companion to the Catechism,* San Francisco: Ignatius Press, 376).

In chapter 44 we read:

Our apostles, too, were given to understand by our Lord Jesus Christ that the office of the bishop would give rise to intrigues. For this reason, equipped as they were with perfect foreknowledge, they appointed the men mentioned before, and afterwards laid down a rule once for all to this effect: when these men die, other approved men shall succeed to their sacred ministry.... Happy the presbyters [priests] who have before now completed life's journey and taken their departure in mature age and laden with fruit (*Epistle to the Corinthians,* 44, quoted in *Companion,* 377)!

In chapter 40 of Clement's letter we see the distinction between the priesthood of all believers and the ordained priesthood:

...[T]o the priests a proper place is appointed, and...[the] layman is bound by the ordinances of the laity (*Epistle to the Corinthians,* 40, trans. Jurgens).

This is likewise seen in the second century *Dascalia Apostolorum:*

In your holy churches, your assemblies, arrange places for the brethren carefully and with all sobriety. Let a place be reserved for the presbyters [priests] in the midst of the eastern part of the house, and let the throne of the bishop be placed among them; let the presbyters [priests] sit with him; but also at the other eastern side of the house let the laymen sit; for thus it is required that the presbyters [priests] should sit at the eastern side of the house with the bishops, and afterwards the laymen, and then the women. When we pray the bishop and presbyters should stand first, followed by the laymen and women (2;8;9).

Ignatius of Antioch, the disciple of the apostle John, reaffirms Clement and the *Dascalia* with the following statement:

Every man who belongs to God and Jesus Christ stands by his bishop.... (Philadelphia, 3, trans. Jurgens). Follow your bishop, every one of you, as obediently as Jesus Christ followed the Father. Obey your clergy [priests] too, as you would the apostles; give your deacons the same reverence that you would to a command from God. Make sure that no step affecting the church is ever taken by anyone without the bishop's sanction. The sole Eucharist you should consider valid is one that is celebrated by the bishop himself, or by some person [presbyter/priest] authorized by him. Where the bishop is to be seen, there let all his people be; just as wherever Jesus Christ is present, we have the catholic Church.... This is the way to make certain of the soundness and validity of anything you do.... (Smyrnaeans, 8). Let the bishop preside in the place of God, and his clergy [presbyters/priests] in the place of the apostolic conclave, and let my special friends the deacons be entrusted with the service of Jesus Christ.... (Magnesians, 6). Without these three orders no church has any right to the name (Trallians, 3).

The gift of the priesthood is exactly that, a magnificent gift of God's love for his people.

St. Francis of Assisi once gave an account of what he would do if he was approached by an angel on one side and the most evil priest to have ever existed on the other. He asked his confreres which one would they give respect to, the angel or the evil priest? They all responded by saying, *"Obviously the angel. We would want nothing to do with the evil priest!"* Francis responded by saying that he would give respect to the most evil priest first, for the priest can do what the angel cannot. The priest can bring Jesus to him in the Eucharist, Jesus' Body, Blood, Soul, and Divinity. The priest can forgive his sins, the angel cannot! The priest can make present Calvary! The disciples of Francis now understood the great gift of the priesthood.

Why celibate priests?

Often people argue that it is unfair that priests cannot marry. Some even go so far as to say that an unmarried priesthood is evil and contrary to God's will, for Genesis reminds us to "be fruitful and multiply."

Why do Catholics of the Roman rite have unmarried priests?

God's providence has for many reasons shown the beauty and dignity of a celibate priesthood. From a practical point of view (a view we often want to overlook for various reasons, including discomfort) if we had married priests we would as a community have to pay and support the homes and the families of these married priests. That means health insurance and a pension plan for the wife and kids. It means college tuition for the children when they grow up. If the individual parish communities had to pay for all priests that were married, the Church would suffer greatly from a financial point of view. In fact it would be disastrous. How many poor people would go without food? How many people that are served by all aspects of Catholic charitable organizations would suffer because of money being directed toward the families of priests as opposed to others? Given this, you can just imagine the strain the Church would be placed in—particularly when you take into account the fact that Catholics are notorious for not tithing.

A second reason that points to the impracticality of a married priesthood can be found in those members of other denominations that have converted to the Catholic faith and have received permission from Rome to become priests in the Roman Church while keeping their families together. Many such wives of these married priests mention that they can see how much time is taken away from the parish community by their husbands' family responsibilities. After all, a husband has an obligation to his wife and children. That is his primary vocation (cf. 1 Cor. 7:32-35).

An unmarried priest is freed from such an obligation and therefore is free to serve the people at all times, including at three o'clock in the morning when an emergency call comes in.

A fourth point is that the priest is, in the words of Pope John Paul II, a "sign of contradiction" in a world drenched in promiscuity. It reminds the faithful that the gift of sex is precious. It is a gift to be experienced in a bonding, unitive way, with an openness to life, and with a respect to the natural order.

Finally, the most important reason for a celibate priesthood comes from Christ himself. Let us remember that Jesus was not married, and that a priest acts in "the person of Christ the Head," and as "another Christ" (cf. Mt. 18:18-19; 2 Cor. 5:18-20). Therefore, Jesus sees celibacy as a gift for the sake of the kingdom of God (Mt. 19:12): *"Some are incapable of marriage because they were born so; some, because they were made so by others; some, because they have renounced marriage for the sake of the kingdom of heaven. Whoever can accept this ought to accept it."*

We see the gift of celibacy and the blessedness of celibacy in Paul's writings in the Scriptures. Paul himself was celibate (cf. 1 Cor. 7:8). Let us reflect on 1 Corinthians 7:32-35.

I should like you to be free of anxieties. An unmarried man is anxious about the things of the Lord, how he may please the Lord. But a married man is anxious about the things of the world, how he may please his wife, and he is divided.

Virginity and celibacy are also commended in 1 Corinthians 7:8-9, 36-40, and 1 Timothy 5:9-12. In Revelation 14:3-4 the 144,000 saved (a symbolic number) are described as virgins. And in Matthew 22:30-32 and

Mark 12: 25-27 we are reminded that those in "heaven neither marry nor are given in marriage." The priest, as a sign of contradiction, is a reminder of what our ultimate future in heaven will be like.

Given what has been said, the Church does recognize that celibacy is a discipline and not a doctrine of the faith. That is why many Catholics from the Eastern rites of the Church do in fact marry.

The Roman rite has chosen to keep the practice of celibacy. While it is true that many of the early popes and bishops were married, and in fact most of the apostles were married, with the exception of Paul and John, the Church has always seen two currents of priestly life, one which incorporated celibacy into the priesthood, and one which incorporated marriage into the priestly life. Both are currents that have existed from the beginning, and both are precious to the Church. Ignatius of Antioch (ca. 107) reminds his clergy in his letter to Polycarp (5): "If anyone can live in a celibate state for the honor of the Lord's flesh, let him do so without ever boasting." Tertullian (ca. 200) in *The Demurrer Against the Heretics* (40, 5) states that the Lord has "virgins and celibates" in his service.

Why "Father"?

Why do we call priests "Fathers"? The Bible says that we are to call no man "Father" (Mt. 23:9). Are Catholics being disobedient to God by calling their priests "Father" and are priests promoting this disobedience by allowing themselves to be called "Father"?

A priest is referred to as "Father" because the early apostles referred to themselves as "Fathers." When we look at Paul's First Letter to the Corinthians (1 Cor. 4:15-17) and John's Epistle (1 Jn. 2:12f) we see that these two apostles perceive and name themselves as "Fathers." Right in the Bible we have two apostles referring to themselves as "Fathers." As the apostle Paul states: "I became your father in Christ Jesus through the gospel" (1 Cor. 4:15). And in 1 Corinthians 17 the apostle Paul refers to his friend Timothy as his "beloved and faithful son in the Lord."

Paul never shied away from referring to others and himself as a father. In Acts 22:1 Paul addresses the Jerusalem Jews as "brothers and fathers." In Romans 4:16-17 Paul calls Abraham "the father of us all." In 1 Thessalonians 2:11 Paul reminds the Thessalonians that he has "treated each one as a father treats his children," and in 1 Timothy 1:2 and Titus 1:4 he calls Timothy, "my true child in faith" and Titus "my true child in our common faith." In Paul's letter to Philemon 1:10 he encourages the community to accept Onesimus when he states: "I urge you on behalf of my child Onesimus, whose father I have become in my imprisonment, who was once useless to you and me but is now useful to both you and me."

In Acts 6:14 and 7:2 Stephen, the first martyr of the Church, calls the Jewish leaders "fathers." And in Hebrews 12:7-9 we are reminded that we have earthly "fathers" to discipline us.

Is this a contradiction? No. Jesus in Matthew 23:9 is pointing out that we have one ultimate Father, one ultimate source of being and teaching. God is the ultimate Father. He is also pointing out that the title "Father" can be abused when the person who bears the title does not bear it worthily.

Paul and John are not pointing to the same understanding of "Father" as is seen in Matthew's Gospel. They are primarily pointing to a spiritual fatherhood in the sense of spiritual guides who proclaim the Gospel by their lives and works.

Christ placed Paul, John and all the apostles as spiritual guides to the ultimate Father, God. In turn, all those with authentic authority may bear the name of "Father" as understood by Paul and John. Thus, priests, by means of the gift of Holy Orders, serve as spiritual guides for their communities. They serve as spiritual "Fathers."

Jesus himself uses the term "father" in Matthew (15:4-5; 19:5, 19, 29; 21:31), John (8:56) and several other places. Jesus actually has Lazarus using the address "father Abraham" twice (Lk. 16:24, 30). In Acts 7:38-39, Acts 7:44-45, and Acts 7:51-53 "father Abraham"—as mentioned before—is attested to as our father in faith.

Ironically, Matthew 23:9 also mentions that we are not to call anyone "teacher." Yet ministers often call themselves "doctor" which is the Latin for "teacher."

Anti-Catholics, in an effort to avoid the name "father," will often address priests by the title "Sir." This is ironic since Jesus is never referred to as "father" but many times as "Sir."

Why do Catholics not allow women priests?

The issue of women priests is not really a matter of allowing or not allowing women to be admitted to the priesthood as much as it is a call to be obedient to the deposit of faith. Christ, his apostles, and all their successors, never ordained women to the priesthood. Two thousand years of Sacred Tradition cannot be wrong.

Some may argue that Jesus and his apostles were living in a paternalistic society and therefore were caught up in the culture of their time which viewed women less than kindly.

The reality is, however, that although Christ was influenced by his culture as a man, he was not bound nor controlled by his culture. After all, Jesus was and is the Son of God. He walked the earth as God and man, as fully divine, fully human—knowing all that was necessary for our salvation in faith and morals. Such a mystery cannot be bound to culture.

Jesus' very life illustrates this. He associated with sinners and had a great many faithful women followers. In one of the Gospels the resurrected Jesus appears to the women first! Furthermore, with the exception of John, it was the women who stood at the foot of the cross; the other apostles had all run away.

The ordination of women is an issue of faith, and as mentioned above, Jesus, despite growing in wisdom and understanding (cf. Lk. 2:52), knew all that was necessary for our salvation (CCC 474; cf. Mk. 8:31; 9:31; 10:33f; 14:18-20, 26-30), and priests are necessary for our salvation (cf. Acts 6:6; 13:3; 14:22; 20:28; Lk. 22:19; 1 Tim. 4:14; 2 Tim. 1:6; Tit. 1:5). Therefore, a belief must have, either in an implicit or explicit manner, indications of existence in the deposit of the faith. Yet, neither the Scriptures nor Sacred Tradition affirm implicitly or explicitly the ordination of women.

Jesus never chose any women to be apostles or their successors. (There are no lines of women successions.) In many ways he chose the least qualified to be his apostles. If qualifications were important, he would have chosen the greatest creature of all time, the Blessed Mother, to serve as an apostle. Yet he didn't choose her to be an apostle.

Women have been granted many gifts. The Church could not go on without them. In virtually all Catholic parishes and Catholic schools, the vast majority of the staff is made up of women. The vast majority of ministries are led by women. The Catholic hospitals and Catholic schools were built on the hard labor and sacrifice of thousands of religious nuns. The greatest Catholic television network, EWTN, is produced and operated by religious sisters. Catholicism in this country and in all countries owes a great deal to the response of women to the gifts of the Spirit.

Yet in God's divine plan, just as men have not been given the gift of childbirth, so women have not been given the gift of Holy Orders. Just as some men are given the gift of living a married life, some men are given a gift of living a priestly life. We could go on and on. The point is that we all have a part to play in God's divine plan, in Christ's Body, the Church.

Some women may argue that they feel "called" to the priesthood by God. The feeling of being called does not mean that one is authentically called. I "felt" called to the priesthood, but it was not until that call was confirmed as authentic by the "laying on of the hands" by a bishop that I knew for certain that in God's providential will I was to serve his people as a priest.

The priest is a man of the Church. The priest is, as what has been previously mentioned, "another Christ" or a person who "acts in the Person of Christ the Head" (cf. Mt. 18:18-19; 2 Cor. 5:18-20). Jesus was a man, and so to best fulfill this imagery of acting in his place he chose men.

Men and women are equal, yet God has chosen them to serve his Church in different and indispensable ways. The following quotes from the Scriptures might be helpful in determining the Church's constant Tradition and understanding regarding the beauty and role of women and the Church's position on the ordination of women: Genesis 2:22; 12:4; Song of Songs 1:8; 4:1-5; 7:2-10; Proverbs 19:14; Sirach 7:19; 26:14f; 36:22-24; Matthew 19:4; Mark 10:7f; 1 Corinthians 11:7; 14: 34f, 37; 1 Timothy 2:11f.

The ordination of women is not in the "deposit of the faith." No matter how politically correct or appealing something may be, if it is not in the deposit of the faith, the Church cannot do anything to change that which is in this deposit! In the *Catechism of the Catholic Church*, which was approved by the universal Church, the bishops in union with the Holy Father—and therefore infallible by means of the ordinary magisterium of the Church in this particular teaching—states: *"The Church recognizes herself bound by this choice made by the Lord himself. For this reason the ordination of women is not possible"* (CCC 1577).

Is there such a thing as the Last Rites?

In the Bible, in James' letter, chapter 5 verses 13-15, we read:

Is anyone among you sick? He should summon the presbyters [priests] of the church, and they should pray over him and anoint [him] with oil in the name of the Lord, and the prayer of faith will save the sick person, and the Lord will raise him up. If he has committed any sins, he will be forgiven.

This is the Sacrament of the Anointing of the Sick or what was often called the "last rites." Origen writing in 244 AD affirms this biblical teaching when he wrote:

Let the priests impose their hands on the sick and anoint them with oil and the sacrament will heal the sick persons and forgive them their sins.

                    Homilies on Leviticus, 2, 7, 8, trans. Jurgens

The Sacrament of Anointing confers a special grace on those suffering from illness or old age. It is a sacrament that can only be administered by a bishop or a priest.

Its power is in the unifying of a person's sufferings with the Passion of Christ. It brings God's healing and loving presence upon the person.

At times the healing is spiritual, at times it is emotional or physical, but God brings about in the person whatever is best for a person's eternal destiny, his or her salvation (cf. Jms. 5:13f).

If a person is unable to receive the Sacrament of Penance—for example, due to incapacitation—the sacrament forgives the sins of the person (Mk. 6:12-13; Jms. 5:13-15).

Prayer and the Doctor?

A small number of groups feel that if one's faith is strong enough God will heal them of any disease. Many television faith healers are so successful it is a wonder that anyone with faith should ever become ill, or for that matter, ever die.

We as Catholics believe that God does in fact heal people, but we also believe that God has given us the gift of doctors whom God has blessed to help people in their ailments.

Prayer is a must, but so is a good doctor and good medicine. In fact, many miracles happen in cooperation with a doctor's medical attention.

Let us never forget Paul's advice to Timothy: "Stop drinking only water, but have a little wine for the sake of your stomach and your frequent illnesses" (1 Tim. 5:23). In other words, take something for your stomach ailments and illnesses.

What about blood transfusions?

Because the "eating" or "drinking" of blood is forbidden by God in Deuteronomy 12:23-25, many Jehovah's Witnesses prohibit the medical practice of blood transfusions (see also Gen. 9:3-4; Lev. 17:14; Acts 15:29).

How do we answer the Jehovah's Witnesses?

First of all, the prohibition is against the "eating" and "drinking" of blood, not of transfusing blood into one's veins.

Secondly, the prohibition against the "eating" and "drinking" of blood was primarily meant to provide a manner of separation between the beliefs of the people of God and the pagans. It was also meant to keep the people of God healthy.

By the time of Jesus, these realities would forever change: Peter is reminded that God has made all things clean (Acts 10:9-16) and Jesus reminds us that "unless you eat the flesh of the Son of Man and drink his blood, you do not have life within you. Whoever eats my flesh and drinks my blood has eternal life, and I will raise him on the last day. For my flesh is true food, and my blood is true drink. Whoever eats my flesh and drinks my blood remains in me and I in him" (Jn. 6:53-56).

The transfusion of blood is in no way associated with paganism nor dietary contamination. For that matter, it is not even biblical.

Is marriage a sacrament?

How can I ever express the happiness of a marriage joined by the Church, strengthened by an offering, sealed by a blessing, announced by angels, and ratified by the Father? How wonderful the bond between two believers, now one in hope, one in desire, one in discipline, one in the same service! They are both children of one Father and servants of the same Master, undivided in spirit and flesh, truly two in one flesh. Where the flesh is one, one also is the spirit. Tertullian (ca. 155-240)
Ad uxorem, 2, 8, 6-7: PL 1, 1412-1413

Jesus infused his very presence into the wedding feast at Cana (Jn. 2:1f) and forever changed the mystery of marriage.

In Matthew 19:5-6 we read: "A man shall leave his father and mother and be joined to his wife, and the two shall become one flesh. They are no longer two, but one flesh. What God has joined together, no human being must separate." Just as Christ's union with his Body, the Church, cannot be separated (Eph. 5:22-32), likewise the union between husband and wife, a union which mirrors the relationship between Christ and his Church (ibid.), cannot be separated.

Christ elevated marriage to the level of a sacrament by the gift of grace, the gift of his very self. The reality of a man who gives himself completely, without doubt, without reservation, fully to his wife, and a wife who gives herself completely, without doubt, and fully to her husband can only come about by the supernatural gift of grace. It is only in this way that two can really become one (Mt. 19:3-6; Mk. 10:6-9).

Because of this unity to which God calls a couple, marriage must be holy, indissoluble, open to life, and according to the natural order (Mt. 19:5; Mk. 10:7f; Eph. 5:22-32; 1 Thess. 4:4; 1 Tim. 2:15; Gn. 38:9-10; Lv. 20:13). Marriage must mirror Christ's love for his own Bride, the Church (Eph. 5:25, 31-32). It must mirror God's covenant with his people (cf. Song of Songs).

Because of the above reality, marriage is that which must be blessed by the Church. As Ignatius of Antioch (ca. 107), the disciple of the apostle John, states:

> *It is right for men and women who marry to be united with the bishop's approval. In that way their marriage will follow God's will and not the prompting of lust (Letter to Polycarp, 5).*

Marriage is that precious gift where spouses are called to aid each other on the journey towards holiness. Marriage is a vocation directed toward the salvation of spouses and the perpetuation of the mystery of Christ and his Church to the world.

What is an annulment?

People often refer to annulments as the Catholic version of divorce. Nothing could be farther from the truth!

A Catholic annulment does not deny that a civil, worldly or paper marriage existed. But what an annulment does assert is that this civil union was not a sacramental union, a marriage elevated by God's blessing. In other words, it was a civil marriage that was never elevated to the level of a sacramental marriage.

How can this be? The answer lies in what makes a sacramental marriage: The key to a valid sacramental marriage is based on consent. Two people must enter into marriage freely and without any natural (i.e., pathological or psychological) or ecclesiastical hindrance (i.e., outside the proper form required by the Church).

It is important to recognize that a marriage under its proper form is always presumed to be sacramental, no matter what pathological or psychological factors may be present in the marriage. If a couple remains together, grace is keeping it together in all likelihood. However, if at one point the marriage breaks up, then the Church can investigate, upon the request of a spouse or spouses, whether the consent at the time of the wedding was possibly invalid, whether a couple or one of the spouses lacked the capacity for making a true and valid consent.

The determination of the validity of the consent between spouses at the time of their wedding is left to professionals in various fields, including canon lawyers and judges. Some of the common grounds for annulments are as follows:

- Inability to be other-oriented, to engage in interpersonal relationships
- Hidden lies or lack of honesty in one's identity or personality
- Lack of psychological or emotional maturity
- Alcoholism, drug addiction
- Sexual dysfunction
- Lack of desire for children
- Lack of desire to raise one's children as Catholics
- Inability for faithfulness
- Inability to comprehend the permanence of marriage
- Pressure to marry

Of course these are but a few of the grounds for annulments. And again, I must strongly and fervently reiterate that a marriage that follows the proper ecclesiastical form is **always presumed to be valid unless challenged!** For example, a spouse may be married to a drug addicted, alcoholic, manic-depressive spouse, but if that bond between the husband and wife remains intact, then one must assume that the grace of God is giving this couple the capacity to deal with these difficulties—and therefore the marriage is assumed to be sacramental.

But one might ask how do we justify an annulment in terms of Scripture? After all, doesn't the Bible say: "A man shall leave his father and mother, and he shall cling to his wife, and the two shall become one" (Eph. 5:31); "they are no longer two but one flesh" (Mk. 10:8); and "what God has joined no man must separate" (Mk. 10:9; cf. 16:18; 1 Cor. 7:10-11).

This is where philosophy serves us well. If there are marriages that God has joined together, there must necessarily be some marriages or unions which God has not joined together. Likewise, the reality of two becoming one in marriage implies that one must in fact have the free will and capacity to live this reality of oneness! Hence, from a purely philosophical point of view, some marriages are not sacramental marriages, that is, marriages elevated to the level of a sacrament since they are not joined by God's blessing nor are they blessed with the ability of two people becoming one.

Scripture supports these philosophical conclusions when it refers to "unlawful marriages," marriages prohibited by God (cf. Acts 15:20; 15:29; Mt. 19:5-9; cf. Lev. 18).

It is in part for this reason that John the Baptist was beheaded. John condemned the unlawful, invalid relationship between King Herod and Herodias, the wife of Herod's brother Philip (Mt. 14:3-12).

An annulment is a recognition of a non-binding, non-sacramental union. It is based on Scripture and the natural philosophical conclusions that flow from the Scriptures.

IV
THE TRINITY AND THE COMMUNION OF SAINTS

Is there a Trinity? Who is Christ? Who is the Holy Spirit?

Christians are brought to future life by one thing...that they recognize that there is a oneness, a unity, a communion between the Son and the Father, and that there is a oneness, a unity, a communion, albeit a distinction, between the Spirit, the Son, and the Father.

Justin Martyr (ca. 148 AD), Legat. Pro Christ

In Genesis 1:26 we read in the story of creation: "Then God said: 'Let *us* make man in *our* image, after *our* likeness." In Genesis 3:22 we read: "Then the Lord God said, 'See, the man has become like one of *us*...'" Who is this *us*? Who is this *our*?

In Genesis 18:1-3 many of the early Fathers of the Church saw the foreshadowing of the Trinity. Abraham meets three mysterious men (understood to be three divine persons or guests) whom he addressed by the singular as opposed to the plural, phrase, "My Lord" (a title most often reserved by the Jews for God). When studying the shifting back and forth between the three divine guests and Yahweh, we find something of great interest. At times Yahweh represents the three, at times Yahweh is one of the three, at times all the three refer to a single divine being. When taken as a whole, we can see why the Church Fathers often saw the foreshadowing of the Trinity in this passage.

In the Hebrew Scriptures, the Old Testament, God is referred to as Yahweh or Elohim, Elohim being translated as God and Yahweh being translated as Lord. What is of particular interest is that Elohim is a plural noun for God. The name Elohim indicates a oneness and a plurality. Therefore, Elohim indicates a oneness and a plurality in God. For the Christian, there is one God in three persons or modes of expression; there is a "oneness," yet a "plurality." The Hebrew Scripture's, the Old Testament's, name for God attests to the plurality of persons in the one God!

At the baptism of Jesus in the Jordan we hear: "When Jesus had been baptized he at once came up from the water, and suddenly the heavens opened and he saw the Spirit descending like a dove and coming down on him. And suddenly there was a voice from heaven, 'This is my Son, the Beloved; my favor rests on him'" (Mt. 3:16f). Right at his baptism the Trinity was manifested to the world: The voice is the Father, Jesus is the Son, and the image of the dove is the Holy Spirit.

At the Transfiguration Jesus is manifested in the midst of a cloud casting a shadow: "A bright cloud covered [Moses, Jesus, and Elijah] with a shadow, and from the cloud came the voice which said, 'This is my beloved Son in whom I am well pleased. Listen to him'" (Mt. 17:5). The voice is that of the Father, the beloved one is Jesus, and the cloud covering Jesus with a shadow is the Holy Spirit (Note that throughout the Hebrew Scriptures, particularly in Exodus, Numbers, and Deuteronomy, a cloud or a shadow represents God's presence).

Before Christ's ascension he reminded his apostles to go throughout the world and baptize in the "name of the Father, and of the Son, and of the Holy Spirit" (Mt. 28:10). Notice Jesus says to baptize in the "name" (singular) and not the "names" (plural) "of the Father, and of the Son, and of the Holy Spirit." There is one God, yet three Persons within that one God.

In the Gospel of Luke and the Acts of the Apostles, the sacred writer emphasizes the theme that in the Hebrew Scriptures the Father was most apparent, in the Christian Scriptures, the New Testament, Jesus became the center of attention, and in the acts of the early Church, the presence of the Holy Spirit was the main character. Obviously all three persons of the Trinity are present in each, yet the revelation of the Trinity took place through the process of revelation.

The word "Trinity" is first recorded in the writings of Theophiles, the bishop of Antioch, to describe the mystery of One God in Three Persons, Father, Son, and Holy Spirit. And according to the ecclesiastical writer Tertullian, by the year 211, the signing of oneself with the sign of the cross, with the sign of the Trinity, had become a well-established custom of the Christian faithful.

"Yahweh is the true God and there is no other" (Dt. 4:35). God is One and has One nature (cf. Is. 40:25-28; 43:10-13; 44:6-8; 1 Chron. 17:20; Mk. 12:29; 1 Cor. 8:4-6). There are Three Persons in One God, the Father, the Son, and the Holy Spirit. There is no confusion, change, division, nor separation between the Persons of the Trinity.

The Three Persons are distinct in their relations of origin: The Father generates, the Son is eternally begotten (Jn. 1:1-4f) and the Holy Spirit eternally proceeds from the Father and the Son (cf. Jn. 15:26).

Let us examine the following Scripture citations: John 1:1; 5:18; 10:30; 14:1; 15:26; 16:14; 17:10; 20:28; Acts 5:3f; 13:2; 13:21; 20:22; 20:25; Romans 9:5; Philippians 2:5f; 3:3; 1 Corinthians 2:10; Galatians 5:18; Ephesians 6:18. When we look at and study the following Bible quotes in relation to each other, we come to the incontrovertible conclusion that we have a Triune God.

The Son

I will redeem you with an outstretched arm and with mighty acts of judgment.

God (Exodus 6:6)

At Jesus' birth the prophet Isaiah is quoted (Mt. 1:23: Is. 7:14): "'Behold a virgin shall be with child and bear a son, and they shall name him Emmanuel," which means God is with us.'" Jesus is "God [who] is with us." "In Christ the fullness of deity resides in bodily form" (Col. 2:9). He is the "I AM" of the burning bush (Ex. 3:14) as attested to by the apostle John (Jesus makes use of the words "I AM" in reference to his divinity at least 11 times in John's Gospel alone). Jesus is, as Thomas proclaims, "my Lord and my God" (Jn. 20:28). In the Hebrew Scriptures the title "Lord" referred to Yahweh, to Adonai, to God. In the New Testament, the name "Lord" refers to only one reality, God! In John 20:16 Mary Magdalene calls the risen Christ by the title "Rabbuni" which was often a title used to address God. Likewise, the phrase "to him be glory for ever," which is often used to address Jesus (2 Tim. 4:18; 2 Pet. 3:18; Rev. 1:6; Heb. 13:20-21), is a phrase that was usually reserved to God alone (cf. 1 Chron. 16:38; 29:11; Ps. 103:31; 28:2). And in Titus 3:5 Jesus is referred to as "God our savior." The Old Testament God is likewise referred to as savior: "It is I the Lord; there is no savior but me." The Old Testament "savior" and the New Testament "savior" are one and the same, since there is "no savior but [God]" (Is. 43:11). Can anyone doubt that Jesus is Lord and God? Can anyone doubt that whoever has seen Jesus has seen the Father (Jn. 14:9)?

The Son of God that existed from all eternity became incarnate some 2000 years ago. He assumed a human nature (Mt. 1:21; Lk. 2:7; Jn. 19:25). The Son is one Person, the second Person of the Trinity, with two natures, a human nature (like us in all things but sin) and a divine nature. He is fully human, fully divine, and there is no confusion, change, division, nor separation between the two natures (cf. Mt. 3:17; 9:6; Mk. 1:1; 8:31; Lk. 1:32; 19:10; Jn. 1:34; 3:13-14; 8:46; Rm. 1:3; 2 Cor. 5:21; Heb. 4:15; 1 Pet. 2:22).

The reality that Christ is both human and divine, the God-man, is attested to by Ignatius of Antioch (ca. 107), the convert of the apostle John, and the friend of Peter and Paul when he says that Christ is "both flesh and spirit, born and unborn, God in man, true life in death,…first subject to suffering and then beyond it."

The Governor of Pontus, Pliny, in his *Letters to the Emperor Trajan* (ca. 111-113) makes reference to the common practice of Christians "singing hymns to Christ as God." Likewise, at the Jewish Council of Jamnia (ca. 95) expelled Christians from the synagogues in part because of their belief that Christ was God as well as a man. For the early Christians, Jesus was fully human, fully divine, the God-man.

Jesus had a divine will and a human will. His human will was in complete conformity with his divine will (Mt. 11:25; Lk. 2:49; Jn. 4:34; Phil. 2:8).

Jesus assumed a human, rational soul (Phil. 2:7f). In his human nature

Jesus grew in "wisdom and stature" (Lk. 2:52). In his human nature Jesus had the "fullness of understanding of the eternal plan he had come to reveal" (Mt. 13:32).

Jesus is the Creator of all things (Jn. 1:3; Col. 1:16f; Heb. 1:2), the Lord of Glory (1 Cor. 2:8), the King of kings (Rev. 17:14; 19:16), the Alpha and Omega (Rv.1:7f). He preached the Kingdom (Mt. 3:2; Mk. 1:15; Acts 2:38). He was immune from sin (Jn. 8:46; 2 Cor. 5:21; Heb. 4:15). He died for all (Jn. 3:16f; Heb. 4:15) and rose from the dead (Mt. 12:39f; Acts 1:22; Rom. 4:24; 1 Cor. 14:4) and will come again to judge the living and the dead—a uniquely divine prerogative according to Hebrew theology (Mt. 19:28; 25:31; Jn. 5:22; Acts 10:42). (For a deeper understanding of Jesus in his human and divine natures, review chapter V on Mary, for the "school of Mary" teaches us about the mystery we call Jesus).

The Holy Spirit

"Peter said, 'Ananas, why has Satan filled your heart so that you lied to the Holy Spirit....You have not lied to human beings, but to God'" (Acts 5:3-4).

The Holy Spirit is the Third Person of the Trinity and is the source of Holy works. His divinity and consubstantiality or oneness with the Father and the Son is attested to throughout the New Testament (i.e., Jn. 14:16-18; 14:23; Acts 5:3f; 28:25f; 1 Cor. 2:10f; 3:16; 6:11, 19f; 1 Pet. 1:1-3; Ep. 4:4-6). As the Third Person of the Trinity, He proceeds from the Father and the Son: "When the Advocate comes whom I will send you from the Father, the Spirit of Truth that proceeds from the Father, He will testify to me" (Jn. 15:26).

Through the operation of the Holy Spirit we are made aware of the Incarnation (Mt. 1:28, 20; Lk. 1:35), the mysteries of the Church (1 Cor. 2:10), the forgiveness of sins (Jn. 20:22-23), the justification and sanctification of souls (1 Cor. 6:11; Rom. 15:16), and the charity of God (Rm. 5:5).

The Holy Spirit is the Spirit of truth (Jn. 14:16-17; 15:26). The Spirit strengthens our faith (Acts 6:5), dwells within us (Rom. 8:9-11; 1 Cor. 3:16; 6:19) and guides our works (Acts 8:29). The Spirit gives us a supernatural life (2 Cor. 3:8) with supernatural gifts (1 Cor. 12:11).

The gifts of the Spirit are wisdom, understanding, counsel, fortitude, knowledge, piety, and fear of the Lord (wonder and awe) (cf. Isa. 11:1-2). The fruits of the Spirit are love, joy, peace, patience, kindness, generosity, faithfulness, gentleness, and self-control (Gal. 5:22-23).

Many attempts have been made to understand the mystery of the Trinity. Some attempts to describe the Trinity include that of a married man and father. The man is himself, a husband, and a father. Another analogy is that of the three states of water—water as liquid, as gas, as solid. St. Patrick used the example of the clover by pointing out that there is only one clover, yet three petals. St. Hildegard of Bingen used the example of fire—the flame being made up of a brilliant light, red power, and fiery heat. St. Ignatius of

Antioch (ca. 107) used the example of three notes making one musical sound. St. Gregory Nazianzen (*Orat.* 31:31-32), St. Cyril of Alexandria (*Thesaurus Assert.*, 33), and St. John Damascene (*Fid. Orth.* 1:13) used the image of the sun, the ray, and the light as well as the source, the spring, and the stream.

Despite the fact that the above analogies all fall short of explaining the great mystery of the Trinity, they still help us in striving to grasp that which is beyond our grasp. As Marius Victorinus wrote in *The Generation of the Divine* word in 356 AD:

> *Because no name worthy of God can be found, we give a name to him from those things which we do know, while bearing in mind that we cannot give to God a name or appellation that is proper to him. That is how we say, 'God lives,' or 'God understands.' Hence, from our own actions, we give a name to the actions of God, considering them as being his in a super eminent way; not such as he really is, but as an approach to what he really is. It is likewise in this way that we impose substance, existence, and other such concepts, upon God. And we speak in a certain way of his ousia or essence, in hinting at what really pertains to him and at what his being really is, by the consideration of created substance (28).*

God is mystery. Yet in many ways, this very mystery is what makes the reality of the Trinity all the more true.

We as human beings are attracted to mystery. Thus, since the Trinity is a mystery, we are attracted to it. Even in heaven, while experiencing the beatific vision, there will still be mystery. For as St. Thomas Aquinas states, the very mystery of God in heaven will keep us eternally attracted to him.

"May the grace of the Lord Jesus Christ and the love of God [the Father] and the fellowship of the Holy Spirit be with all of you" (2 Cor. 13:13).

- Pseudo-Christians such as the Mormons often like to make reference to the "gods" in the Old Testament. From their view, they assume these to be real gods, and therefore justify their belief that they too one day will be gods. This, of course, is a perversion of the always held belief by the Jews in Monotheism. When the Old Testament refers to "gods" it is referring to the worthless, empty, and useless worship of pagan idols. Psalm 115:3-8 illustrates how "gods" are to be understood: "But our God is in the heavens: he does whatever he wills. Their idols [gods] are silver and gold, the work of human hands. They have mouths but they cannot speak; they have eyes but they cannot see; they have ears but they cannot hear; they have nostrils but they cannot smell. With their hands they cannot feel; with their feet they cannot walk. No sound comes from their throats. Their makers will come to be like them and so will all who trust in them." Deuteronomy 4:35 makes it quite clear that there is only one God.

- The earliest picture of the Crucifixion is found scribbled in an army officer's quarters on Palatine Hill in Rome in the early 200's. The caption reads: "Alexamenos worships his God."

Jesus is God!

At Jesus' birth the prophet Isaiah is quoted (Mt. 1:23; Is. 7:14): "'Behold a virgin shall be with child and bear a son, and they shall name him Emmanuel," which means "God is with us." Jesus is "God [who] is with us." "In Christ the fullness of deity resides in bodily form" (Col. 2:9). He is the "I AM" of the burning bush (Ex. 3:14) as attested to by the apostle John (Jesus makes use of the words "I AM" in reference to his divinity at least 11 times in John's Gospel alone). Jesus is, as Thomas proclaims, "my Lord and my God" (Jn. 20:28). In the Hebrew Scriptures the title "Lord" referred to Yahweh, to Adonai, to God. In the New Testament, the name "Lord," "Kyrios," refers to only one reality, God! In John 20:16 Mary Magdalene calls the risen Christ by the title "Rabbuni" which was often a title used to address God. Likewise, the phrase "to him be glory for ever," which is often used to address Jesus (2 Tim. 4:18; 2 Pet. 3:18; Rev. 1:6; Heb. 13:20-21), is a phrase that was usually reserved to God alone (cf. 1 Chron. 16:38; 29:11; Ps. 103:31; 28:2). And in Titus 3:5 Jesus is referred to as "God our savior." The Old Testament God is likewise referred to as savior: "It is I the Lord; there is no savior but me." The Old Testament "savior" and the New Testament "savior" are one and the same, since there is "no savior but [God]" (Is. 43:11). Can anyone doubt that Jesus is Lord and God? Can anyone doubt that whoever has seen Jesus has seen the Father (Jn. 14:9)?

In Revelation 1:8 God is referred to as the "Alpha and the Omega." In Revelation 22:13 Jesus is called the "Alpha and the Omega." Therefore, Jesus is God! In Deuteronomy 10:17 of the Old Testament God is referred to as the "Lord of Lords." In Revelation 19:16 Jesus is referred to as the "Lord of lords." Therefore, Jesus is God!

On Mount Sinai Moses was given the law, the word of God, by God to bring down to his people. On the Mount of Olives, Jesus is the Word and gives the law of life, the Word, directly to his people in the Beatitudes. Therefore, Jesus is God!

The Son of God that existed from all eternity became incarnate some 2000 years ago. He assumed a human nature (Mt. 1:21; Lk. 2:7; Jn. 19:25).

The Son is one Person, the second Person of the Trinity, with two natures, a human nature (like us in all things but sin) and a divine nature. He is fully human, fully divine, and there is no confusion, change, division, nor separation between the two natures (cf. Mt. 3:17; 9:6; Mk. 1:1; 8:31; Lk. 1:32; 19:10; Jn. 1:34; 3:13-14; 8:46; Rm. 1:3; 2 Cor. 5:21; Heb. 4:15; 1 Pet. 2:22). Jesus assumed a human, rational soul (Phil. 2:7f). In his human nature Jesus grew in "wisdom and stature" (Lk. 2:52). In his human nature Jesus had the "fullness of understanding of the eternal plan he had come to reveal" (Mt. 13:32).

The two natures of Christ, human and divine, cause grave problems for groups like the Jehovah's Witnesses and the Mormons. It is a concept they do not seem to grasp. The hymn to the Philippians (2:5-7) is essential for understanding Jesus' two natures, and so I quote in full:

> *Have this mind among yourselves, which is yours in Christ Jesus, who, though he was in the form of God, did not count equality with God a thing to be grasped, but emptied himself, taking the form of a servant, being born in the likeness of men. And being found in human form he humbled himself and became obedient unto death, even death on a cross. Therefore God has highly exalted him and bestowed on him the name which is above every name, that at the name of Jesus every knee should bow, in heaven and on earth and under the earth, and every tongue confess that Jesus Christ is Lord, to the glory of God the Father (RSV).*

It is only in this "emptying," this "kenosis" that one can understand Jesus as God and man, as fully human, fully divine, like us in all things but sin, being entrusted by the Father with only those things which were necessary for our salvation (i.e., those things pertaining to faith and morals).

In his human nature Jesus grew in "wisdom and stature" (Lk. 2:52). In his human nature he could say the "Father is greater than I" (Jn. 14:28) or that only the Father knows the end of time. But in his divine nature he could say "I AM" (Jn. 8:58; 11 times in John's Gospel alone). In his divine nature he could say I am the way, and the truth, and the life (Jn. 14:6). Only God can say that. Humans say "I know the way, I know the truth, I know life," but they certainly do not say "I am the way, and the truth, and the life, and no one comes to the Father except through me" (Jn. 14:6). In his divine nature he could say "the Father and I are one" (Jn. 10:30). In his divine nature he could forgive sin, a divine prerogative (Mt. 9:6) and judge the living and the dead, another divine prerogative (Mt. 19:28).

The reality that Christ is both human and divine, the God-man, is attested to by Ignatius of Antioch (ca. 107), the convert of the apostle John, and the friend of Peter and Paul when he says that Christ is "both flesh and spirit, born and unborn, God in man, true life in death,...first subject to suffering and then beyond it." And in his letter to the Ephesians (18:2) he wrote: "For our God, Jesus Christ, was conceived by Mary..."

The Governor of Pontus, Pliny, in his *Letters to the Emperor Trajan* (ca. 111-113) makes reference to the common practice of Christians "singing hymns to Christ as God." Likewise, at the Jewish Council of Jamnia (ca. 95) expelled Christians from the synagogues in part because of their belief that Christ was God as well as a man. For the early Christians, Jesus was fully human, fully divine, the God-man.

Jesus had a divine will and a human will. His human will was in complete conformity with his divine will (Mt. 11:25; Lk. 2:49; Jn. 4:34; Phil. 2:8).

Jesus is the Creator of all things (Jn. 1:3; Col. 1:16f; Heb. 1:2), the Lord of Glory (1 Cor. 2:8), the King of kings (Rev. 17:14; 19:16), the Alpha and Omega (Rev.1:7f). He preached the Kingdom (Mt. 3:2; Mk. 1:15; Acts 2:38). He was immune from sin (Jn. 8:46; 2 Cor. 5:21; Heb. 4:15). He died for all (Jn. 3:16f; Heb. 4:15) and rose from the dead (Mt. 12:39f; Acts 1:22; Rom. 4:24; 1 Cor. 14:4) and will come again to judge the living and the dead—a uniquely divine prerogative according to Hebrew theology (Mt. 19:28; 25:31; Jn. 5:22; Acts 10:42). (For a deeper understanding of Jesus in his human and divine natures, review chapter V on Mary, for the "school of Mary" teaches us about the mystery we call Jesus).

Jehovah's Witnesses purposely change John 1:1 from the "Word was God" to "the Word was a god." Because they do not believe in the Trinity they have difficulty in understanding verse 1 which states that "the Word was with God, and the Word was God." For them, the second part of the phrase requires changing to correspond to the first part. In other words, how can you be with God and be God? Therefore, the Jehovah's Witnesses changed the original Greek to conform to their belief system. It is similar to what Martin Luther did in Galatians when he added the word "alone" to faith. In both cases, additions and changes were made that are contrary to the original manuscripts.

If you believe in the Trinity there is no need to manipulate or change the original manuscripts. The Greek is perfectly clear. The "Word," the "Son," is with God the Father and because he is with the Father he is God. He is one with the Father, yet distinct as the Son.

Finally, to accept the Jehovah's Witnesses version of John 1:1 is to make Jesus into a false god since Isaiah 44: 6 makes it quite clear, "there is no God but me."

And what about the Fathers of Early Christianity?

The teachings are unanimous. Ignatius of Antioch (d. 110), the friend of the apostle John, and the bishop of Antioch through ordination by the apostles Peter and Paul, writes at least sixteen times in seven letters that Jesus is God!

The Epistle to Dignetus (ca. 124) writes: "He sent him as God; he sent him as man to men" (7:4). Melito of Sardis (184) states: "Christ is by nature both God and man" (*Peri Pascha*). Justin Martyr (150) states: "Christ, the incarnate Word, is divine" (*1 Apol. 10 & 63*). Irenaeus (ca. 185) states: "Jesus was true man and true God" (AH, IV, 6, 7, ANF, 469). Tertullian (185) states: "[The Son] proceeds from God; and in that procession is generated; so that He is the Son of God, and is called God from unity of substance with God.... Thus Christ is Spirit of Spirit, and God of God, as light of light is kindled...that which has come forth out of God is at once God and the Son of God, and the two are one. In this way also, He is Spirit of Spirit and God of God... (*Apology* 21). In *On Flesh of Christ* (5) Jesus "was God crucified." Clement of Alexandria (210) wrote that Jesus was "both God and man." (*Exhortation to the Heathen*, 1).

God, not gods!

Pseudo-Christians such as the Mormons often like to make reference to the "gods" in the Old Testament. They like to cite Psalm 82:6 or Genesis 1:26 as proof. From their view, they assume these to be real gods, and therefore justify their belief that they too one day will be gods through a process of "exaltation." Well, we all know what happened to those who tried to become gods (Genesis). This, of course, is a perversion of the always held belief of the Jews in Monotheism.

Genesis 1:26 is proof of the Trinity. One God in three distinct persons: "God created man in his image...(Gen. 1: 27). "Let us make man in our image, after our likeness" (Gen. 1:26). Verse 27 is singular and points to the oneness of God. Verse 25 makes use of the words "us" and "our" to point out the distinction within the oneness. The Hebrew word for God is Elohim, which likewise points to a oneness, yet a plurality or distinction within the oneness.

In terms of Psalm 82:6 the "gods" being referenced here are not even pagan gods. They are evil judges that have misused and abused their God-given authority. The judges of Israel, since they exercised the divine prerogative to judge (Dt. 1:17) were called "gods" even though they were mortals (see Exodus 21:6).

When the Old Testament refers to "gods" it is referring to the worthless, empty, and useless worship of pagan idols. Psalm 115:3-8 illustrates how "gods" are to be understood: *"But our God is in the heavens: he does whatever he wills. Their idols [gods] are silver and gold, the work of human hands. They have mouths but they cannot speak; they have eyes but they cannot see; they have ears but they cannot hear; they have nostrils but they cannot smell. With their hands they cannot feel; with their feet they cannot walk. No sound comes from their throats. Their makers will come to be like them and so will all who trust in them."*

Furthermore, when we examine the original manuscripts, in the Hebrew, we see that when God is being referred to the following pattern is used: *Singular verb + Elohim= God*. When a false god is being referred to the following pattern is used: *Plural verb + Elohim= false gods*. In the Greek Septuagint you have for the real God, *theos,*and for the false gods, *theoi*.

It is important to recognize that at the time of the writing of the Bible atheism was not an intellectual movement. Atheism as we know it today is relatively new to the modern age. So for people of the ancient world the question was whether there was one God or many gods. Revelation teaches us that there is only one true God and that those who worship "gods" are worshipping idols, worshipping nothing that exists!

The Scriptures are very clear there is but one God, as Deuteronomy 4:35 states: "All this you were allowed to see that you might know the Lord is God and there is no other." Deuteronomy 6:4 states: "Hear, O Israel! The LORD is our God, the LORD alone!" Deuteronomy 32:39 states: "Learn then that I, I alone, am God, and there is no god beside me." Isaiah 45:5-6 states: "I am the LORD and there is no other, there is no God beside me... [Toward] the rising and setting of the sun, men may know that there is none besides me." In Isaiah 43:10 we read: "Before me no god was formed, and after me there shall be none." 1 Corinthians 8:4 states: "[There] is one God, the Father through whom all things are and for whom we exist..." In 1 Timothy 2:5 we read: "For there is one God." Ephesians 4:5 states: "[There] is one Lord, one faith, one baptism; one God and Father of all, who is over all and through all and in all." In John 17:3 we read: "Now this is eternal life, that they should know you, the only true God." The list goes on and on. There is only one God! To accept that there is more than one God is to deny the very Scriptures themselves. To believe in "gods" or to want to become "gods" is to place oneself in opposition to the only true God (Gen. 20:2-3).

All the Fathers of the church have affirmed the reality in one and only one God!

Do Catholics worship saints?

Catholics do not worship saints. We honor them or what we as Catholics like to say is that we venerate them. We give a lower form of veneration, called *dulia*, to saints and angels and to Mary we use the term *hyperdulia* to indicate a higher form of veneration. But God alone receives worship or adoration, *latria*.

We as Catholics venerate or honor the saints, but we do not worship the saints. Only God is worthy of worship (Mt. 4:10; Lk. 4:8; Acts 10:26). If we can honor our mother and father (Ex. 20:12), why can we not honor the saints? Peter, James, and John venerated Jesus, Elijah and Moses in the event of the Transfiguration (Mk: 9:4). Joshua fell prostrate before an angel (Jos. 5:14), Daniel fell prostrate before the angel Gabriel (Dan. 8:17), Tobiah and

Tobit fell to the ground before the angel Raphael (Tob. 12:16). If these great ones could venerate angels and saints, why can't we?

- Often, in some English speaking countries (i.e., England), worship and veneration are sometimes used interchangeably. But the Catholic faith has always made a distinction between the honor given to the saints and Mary, and the honor given to God. In the United States we make the distinction between veneration and worship.

What about the communion of the saints?

Let us not forget those who have died in our prayer. Let us not forget the patriarchs, prophets, apostles, and martyrs who bring our petitions to God; let us not forget the holy fathers and bishops who have died as well as all those most close to us who bring our petitions to God.

Cyril of Jerusalem (ca. 350)
Catechetical Lectures, 23 [Mystagogic 5], 90

We as Catholics venerate or honor the saints, but we do not worship the saints. Only God is worthy of worship (Mt. 4:10; Lk. 4:8; Acts 10:26). If we can honor our mother and father (Ex. 20:12), why can we not honor the saints? Peter, James, and John venerated Jesus, Elijah and Moses in the event of the Transfiguration (Mk: 9:4). Joshua fell prostrate before an angel (Jos. 5:14), Daniel fell prostrate before the angel Gabriel (Dan. 8:17), Tobiah and Tobit fell to the ground before the angel Raphael (Tob. 12:16). If these great ones could venerate angels and saints, why can't we?

We recognize there is only one mediator, Jesus Christ (1 Tim. 2:5). We recognize that Christ is the one mediator, but that he has gifted us and the saints with the ability to engage ourselves in that one mediation. As Paul states: "Be imitators of me, as I am of Christ" (1 Cor. 11:1; also 1 Thess. 1:6-7; 2 Thess. 3:7) In other words, do what I do as I do what Christ does. Isn't this serving in Christ's mediation? Likewise, 1 Thessalonians 1:5-8 reminds us that we must become examples to all believers, and Hebrews 13:7 reminds us that we are to remember our leaders, and that we are to consider and imitate their faith and life. By being a Christian, by being an example of Christ, one shares in Christ's mediation.

Paul also reminds us that "we make up what is lacking in the sufferings of Christ" (Col. 1:24). If this is so, then to be a Christian means that we are by nature sharers in Christ's one mediation.

The very nature of being a Christian is to be a mediator, for by growing in the image and likeness of Christ, which is what it means to grow in holiness, is to inevitably share in Christ's suffering, and by sharing in his suffering means that one shares in Jesus' living sacrifice on the cross to the Father. Just as Christ suffered on the cross for our salvation, we by our suffering, and by being in the image and likeness of Christ, are inevitably participators in the redemptive work of Christ. The lives of the faithful are a living sacrifice to God.

Scripture points out that the saints are first and foremost in heaven with Christ before the general resurrection (2 Macc. 15:11-16; Mk. 12:26-27; Lk. 23: 43; 2 Cor. 5:1, 6-9; Phil. 1:23-25; Rev. 4:4; 6:9; 7:9; 14:1; 19:1, 4-6). God is the God of the living, and not the dead (Mk. 12:26-27). The thief on the cross turns to Jesus, repents, and is reminded that he will be in paradise with him that very day (Lk. 23:43). In Hebrews 12:1 we are reminded that we are surrounded by a cloud of heavenly witnesses. The Old and New Testaments remind us that the martyrs are in the hand of God (Rev. 6:9-11; 20:4; Wis. 3:1-6). The *Didache* affirms: "The Lord will come and all his saints with him."

The Scriptures point to the fact that the faithful on earth are in communion with the saints of heaven (1 Cor. 12:26; Heb. 12:22-24), and that they assist us by their intercessory prayers (Lk. 16:9; 1 Cor. 12:20f; Rev. 5:8). For example, the Scriptures point out that "in his life [Elisha] performed wonders, and after death, marvelous deeds" (Sir. 48:14). Even after death, Elisha was interceding for us and bringing us "marvelous" things. In Tobit 12:12 we see how an angel offers the prayers of the holy ones to God. In Revelation 5:8 we read: "Each of the elders [in heaven] held a harp and gold bowls filled with incense, which are the prayers of the holy ones [being brought to God].

The communion of saints is one of the most precious gifts that God has given us (cf. 1 Cor. 12:24-27). How sad it is for me as a Catholic priest to hear words like "until we meet again" from people of other denominations. For the Catholic, our relationships never end. The communion we share with each other here on earth (1 Cor. 12:24-27) is one that extends into purgatory and heaven. Our relationships change, but they continue into eternity. How comforting to know we are able to help people by our prayers when they are being purified (2 Macc. 12:45). How comforting it is to know that from heaven they are interceding for us in the presence of God (cf. Rev. 5:8; 1 Cor. 12:20f; Heb. 12:22f).

I think of my father who died some twenty years ago. Just as he loved me, cared for me, prayed for me on his earthly journey, what do you think he is doing in heaven? He is loving me, caring for me, and praying for me. But now his prayers are much more powerful, for they are the prayers of a man that has been purified and perfected. They are the prayers of my father at his very best. So when I am having a hard day, I can pray to my father and say, "Hey Dad, say a little prayer to God for me." And he will. Or I can say when

I am having a great day, "Hey Dad, say a little thank you to God for me." And he will. In many ways my father is closer to me now then he ever was before. What a precious gift!

Let us never be fearful of invoking the saints of heaven, for they are a gift that God has entrusted the world with. How many cancerous tumors have disappeared by the invocations of saints? How many ills have been healed by the invocation of the saints? History attests to the miraculous intercession of the saints and their communion with us.

I encourage you to investigate the historical accounts of the canonized saints and the process of canonization. I also encourage you to investigate the accounts regarding the Marian apparitions with particular attention to Lourdes and Fatima. Our God is not a God of the dead, but the God of the living (Mt. 22:32; Mk. 12:27).

Christ is the One True Mediator, but we and the saints in communion with us have been gifted with sharing in that one mediation.

Furthermore, may we never forget the greatest saint of all, Mary. At the wedding feast of Cana it was Mary who interceded for the wedding couple in order for them to have more wine. Jesus performed his first miracle, the turning of water into wine, for his mother (Jn. 2:1-11).

May we remain in communion with God and all his saints, for to love and venerate the saints is to honor God (cf. Gal. 1:24), for his saints are the beauty of his creation and will.

The communion of saints is a sign of the reality of the Trinity. All of Creation echoes the image of the Trinity. Since the Trinity is a communion of Persons, a communion of love (Gen. 1:26), it only makes sense that what he created in his image and likeness (Gen. 1:27) would engage in a similar communion.

Saints and angels, because of their union with God, are worthy of veneration for they reflect their maker. As 1 John 3:2 explains, "We shall be like him, for we shall see him as he is." If this is so, then saints are worthy of respect and veneration.

Jesus, the fulfillment of Judaism's hopes!

In Matthew 1:1f and Luke 1:32-33; 3:33 we are reminded that the Messiah, the "blessed one," the "anointed one," the "savior of his people" would be a descendent of Abraham, Isaac, and Jacob, that he would be of the throne of David and of the tribe of Judah (cf. Gen. 12:3; 17:19; 18:18; 49:10; Nb. 24:17; Is. 9:7). Micah 5:2 makes mention that the Messiah would be born in Bethlehem (cf. Mt. 2:1; Lk. 2:4-7) the town of David. In Isaiah 7:14 we are reminded that from the "house of David" a savior would be born of a virgin and he would be Emmanuel—God is with us (cf. Mt. 1:18, 23; Lk. 1:26-35). His birth would be followed by the slaughter of innocent children (Jer. 31:15; Mt. 2:16-18). He would be forced to flee to and return from

Egypt (Hosea 11:1; Mt. 2:14-15). Born in Bethlehem he would however be known as a Nazarene (Jgs. 13:5; Mt. 2:23).

His ministry would be prepared by John the Baptist, the "voice of one crying in the wilderness" (cf. Is. 40:3; Jn. 1:23). As he grew to maturity, the King of kings, the Lord of lords, would mark the beginning of his "Passion" by a triumphal entry into Jerusalem on a donkey (Zech. 9:9; Jn. 12:13-14). He would be betrayed by his own people (cf. Is. 53:3; Jn. 1:11). He would be betrayed by a friend for 30 pieces of silver (cf. Zech. 11:12; Ps. 41:9; Mk. 14:10). The 30 pieces of silver would be taken and used to buy a "potter's field" (cf. Zech. 11:13f; Mt. 27:6-7). Judas, his betrayer, would be replaced by another, Matthias (cf. Ps. 109:7-8; Acts 1:18-20). The Messiah would be accused by false witnesses (Ps. 27:12; 35:11; Mt. 26:60-61; Mk. 14:57), hated without reason (Ps. 69:4; 35:19; 109:3-5; Jn. 15:24-25), and yet would remain silent (Is. 53:7; Mt. 26:62-63; Mk. 15:4-5). Soldiers would take off, gamble for, and divide his clothing amongst themselves (Ps. 22:18; Mt. 27:35). He would be crucified, pierced in his hands and feet (cf. Zech. 12:10; Ps. 22:16; Mt. 27:35; Jn. 20:27) given gall and vinegar to quench his thirst (cf. Ps. 69:21; Mt. 27:24,48; Jn. 19:19) and placed among thieves on a cross (cf. Is. 53:12; Mk. 15:27-28; Ex. 6:6). Not a single bone would be broken (cf. Ps. 34:20; Ex. 12:46; Jn. 19:32-36). And before his death, he cried out *"Eli, Eli, lama sabachthani"* to remind the faithful that he was the fulfillment of Psalm 22, the innocent, just man who would be sacrificed for the world. To assure he was dead, they would pierce his side (Zech. 12:10; Jn. 19:34). He was pierced for us and our transgressions, taking upon himself the sins of the world (Is. 53:4-5, 6, 12; Ex. 6:6; Mt. 8:16-17; Rm. 4:25; 5: 6-8; 1 Cor. 15:3). He would be placed in a rich man's tomb (cf. Is. 53:9; Mt. 27:57-60). He would be deserted by his followers (Zech. 13:7; Mk. 14:27). He would rise on the third day (cf. Hos. 6:2; Ps. 16:10; 49:15; Lk. 24:6-7; Mk. 16:6-7). He would ascend into heaven (cf. Ps. 68:18; 24:3; Lk. 24:50-51; Acts 1:11; Mk. 16:19) and would be seated at the right hand of the Father (cf. Ps. 110:1; Hb. 1:2-3).

In one of the most poignant moments in the Scriptures, Solomon ponders: "[W]ill God indeed dwell on the earth" (RSV, 1 Kgs. 8:27)? Easter is that unique recognition that God came to dwell among us on earth in the most precious and valued of ways—to teach us to live life abundantly and to live it eternally! Yes, King Solomon, God did come to dwell among us on earth!

V
MARY

Is Mary the Mother of God?

"Look, the virgin shall conceive and bear a son, and they shall call him Emmanuel, which means "God is with us" (Mt. 1:23).

Mary is the mother of Jesus. Mary is the Mother of God. She is the Mother of the Son of God (Lk. 1:35; Gal. 4:4). Who else is this Son of God but God, the second Person of the Trinity?

The Old Testament prefigures Mary as being the Mother of God. During the time of King Solomon until the end of the Kings of Judah, the Queen Mother always sat on the right hand of her son as a confidant and advisor. Jesus, the New Testament Davidic King, has a New Testament Queen Mother, Mary.

Mary is referred to as the "Mother of my Lord" in Luke's Gospel (Lk. 1:43). This is significant for in the Jewish world the title "Lord" was a title reserved for God, Yahweh. And in the Greek New Testament the title "Lord," or "Kyrios" refers to **only** God!! It is no coincidence that just two verses later (v. 45), the divine title "Lord" is being used in such a way that one could not confuse it for anything other than the title for "God." Mary is the Mother of Lord, the Mother of Yahweh, the Mother of God. In the New Testament the title "Lord" refers **only** to God!

Mary is what the ancient Church called the "theotokos," the "God-bearer." In fact, this title for Mary was so common that the anti-Christian emperor Julian the Apostate (361-363 AD) would mock Christians for its "incessant use."

At the Council of Ephesus (431), in seeking to understand more profoundly the mystery of Christ, the Council Fathers could hear the crowds outside the walls chanting "theotokos, theotokos, theotokos!" This was no coincidence. For to truly understand Jesus, the crowds, under the power of the Spirit, knew that one needed to understand Mary.

Thus, it is no coincidence that the identity of Jesus and the identity of Mary would be clarified together. Mary always points to her Son!

The Council Fathers (i.e., bishops) reaffirmed Jesus as being fully human, fully divine without any confusion, division, or separation between his two natures (cf. Mt. 3:17; 9:6; Mk. 1:1; 8:31; Lk. 1:32; 19:10; Jn. 1:34; 3:13-14; 8:46; Rm. 1:3; 2 Cor. 5:21; Heb. 4:15; 1 Pet. 2:22). Mary therefore could not be the Mother of Jesus "only," or the Mother of God "only." To separate Jesus' divinity from his humanity would be to make Jesus into two distinct persons. Yet Jesus is one Person, the Second Person of the Trinity, the Son of God, with two inseparable natures—a human and a divine nature (cf. Mt. 3:17; 9:6; Mk. 1:1; 8:31; Lk. 1:32; 19:10; Jn. 1:34; 3:13-14; 8:46; Rm. 1:3; 2 Cor. 5:21; Heb. 4:15; 1 Pet. 2:22).

All mainline Christians accept this logic, yet why do they not accept the inevitable philosophical conclusion that Mary is—given the above well-accepted Christology—the Mother of God?

Ignatius of Antioch (110 A.D.), the friend of the apostle John, and the bishop of Antioch through the "laying on of hands" by the apostles Peter and Paul wrote the following:

> *Our God, Jesus Christ, was conceived by Mary in accord with God's plan (Ephesians, 18:2).*

The Catholic Irenaeus (180-199 A.D.), the friend of Polycarp, who in turn was the friend of the apostle John wrote:

> *The Virgin Mary...being obedient to His word, received from an angel the glad tidings that she would bear God (Against Heresies, 5,19).*

But what about the three Protestant founders, the three pillars of Protestantism?

Martin Luther, the founder of Protestantism, recognized the important role of Mary as the Mother of God. As he stated in defense of his strong devotion to Mary:

> *Mary was made the **Mother of God**, giving her so many great things that no one could ever grasp them... (The Works of Luther,* Pelikan, Concordia, St. Louis, v. 7, 572).

John Calvin, the second most famous Protestant founder, recognized this reality when he stated:

> *It cannot be denied that God in choosing and destining Mary to be the Mother of his Son, granted her the highest honor.... Elizabeth called Mary the Mother of the Lord because the unity of the person in the two natures of Christ was such that she could have said that the man engendered in the womb of Mary was at the same time the eternal God (Calvini Opera, Corpus Reformatorum,* Braunschweig-Berlin, 1863-1900, v. 45, 348, 35).

And the last of the three fathers of mainline Protestantism, Ulrich Zwingli argues:

It was given to her what belongs to no creature, that in the flesh she should bring forth the Son of God [who is God](Corpus Reformatorum, vol. 6, I).

What ever happened to their Protestant disciples?

Mary is the greatest of God's creatures. She is the greatest creature created by the Son of God.

The greatest cosmic event to ever have occurred, an infinite being, God, becoming, through Mary, a finite being occurred in the Incarnation. How can we deny Mary her special place in Christianity?

Mary is unique for she is the spouse of the Holy Spirit. When we examine the phrase "to overshadow" as used in the annunciation scene in Luke 1:35 we cannot but be made aware of the spousal relationship between Mary and the Holy Spirit. Jewish rabbis knew that the phrase "to overshadow" when used in the context of conception was a euphemism for a spousal relationship (*Midrash Genesis Rabbah* 39:7; *Midrash Ruth Rabbah* 3:9). The Holy Spirit "overshadowed" Mary (cf. Lk. 1:35). Thus, Mary entered a spousal relationship with the Holy Spirit.

No other human being can make the claim of being the "spouse of the Holy Spirit!" And because of this reality, Mary is the Mother of the God who "is with us," (Mt. 1:23). When Mary visited Elizabeth, Elizabeth responded: "And how does this happen to me that the mother of my Lord (Yahweh) should come to me" (Lk. 1:43)? Who is the Lord here in this passage? Jesus. And who is Jesus? God and man, or as Thomas would say, "my Lord (Yahweh) and my God (Elohim)" (Jn. 20:28). (Yahweh was the Yahwistic name for God, and Elohim was the Elohistic term for God. Yahweh is often translated as "Lord" and Elohim as "God"). Mary is the mother of Emmanuel, which means "God with us" (Lk. 1:23). Mary is the Mother of "God with us."

One must study at the "school of Mary," as Pope John Paul the Great explained, if one is to truly understand the mystery which is Christ. No one knows Christ better than Mary. No one can introduce us to a profound knowledge of his mystery better than his mother.

Let us always seek to grasp the mysteries of Mary, for as we do, we will discover the wonders of her Son. As Blessed Mother Teresa would often say, "Let us love Mary as much as Jesus loved her, nothing more, nothing less." Can we ever love Mary as much as Jesus?

Mother of God listen to my petitions; do not disregard us in adversity, but receive us from danger.
Second Century Papyrus, Or. 24, II.

The fact is this:
1) All mainline Christians affirm that Mary is the mother of Jesus.
2) All mainline Christians affirm that Jesus is God.
3) Therefore, when push comes to shove, all mainline Christians, no matter how uncomfortable, recognize that Mary is the Mother of God. To deny this reality is to distort who we all agree Jesus is and who we all agree is the Triune God.

What is the Immaculate Conception?

In the year 306 AD we read in Ephraeim's *Nisbene Hymn* (27, 8) the following:

> *You alone and your Mother*
> *Are more beautiful than any others;*
> *For there is no blemish in you,*
> *Nor any stains or sins upon your Mother.*

Ambrose (340-370) wrote of Mary:

> *Lift me not up from Sara but from Mary, a Virgin not only undefiled but Virgin whom grace has made inviolate, free from every stain of sin (Commentary on Psalm 118, 22, 30).*

The Immaculate Conception is the teaching which affirms that Mary was redeemed by Jesus from the very moment of her conception. She was preserved from "original sin" and personal sin by Jesus, her Redeemer and Savior.

Where do Catholics come up with this teaching? The teaching has always been part of the deposit of faith and can be seen through the logical philosophical implications that flow from Luke 1:28. Mary is *kecharitomene*; that is, she is "full of grace." She is full of grace because of Christ and thus if one is full of grace one cannot have the stain of "original sin" or the stain of any personal sin; otherwise, the angel would have said: "Hail Mary, partially full of grace." But the angel didn't say that as we all know. He said, "Hail Mary, full of grace." Jesus was without sin, and because of Jesus, Mary was without sin.

Mary is the New Eve as Jesus is the New Adam. In the Garden of Eden the devil, a fallen angel, brought the words that would lead to death. At the annunciation, the angel Gabriel would bring the words that lead to life to Mary. Eve disobeyed God and brought upon the fall of the human race. Mary obeyed God and helped to bring about the redemption of the human race. Where Eve was a poor disciple and poor mother, Mary, the New Eve, was the perfect disciple and perfect mother. Jesus, the New Adam, was without sin;

Mary, the New Eve, by virtue of her son, is likewise without sin! Mary, which means "excellence" or "perfection," truly lived up to her name.

As Jesus is the New Adam (1 Cor. 15:45), Mary is the New Eve. After the fall of Adam and Eve we read in Genesis 3:15: "I will put enmity between you and the woman, and between your offspring and hers; He will strike at your head, while you strike at his heel." The woman's son, Jesus, will crush the head of the serpent. If this is so, then who is the woman? Mary! Just as Adam and Eve brought death to the world, the New Adam and Eve, Jesus and Mary, bring life to the world, bring redemption. The Old Adam and Eve became sinners and brought sin into the world; the New Adam, Jesus, and the New Eve, Mary, remain sinless and are responsible for bringing about the redemption in the world.

St. Irenaeus, the friend of Polycarp, who was in turn the friend of John the Apostle wrote:

> *Just as [Eve]...having become disobedient, was made the cause of death for herself and for the whole human race; so also Mary, ...being obedient, was made the cause of salvation for herself and the whole human race.... Thus, the knot of Eve's disobedience was loosed by the obedience of Mary. What the virgin Eve had bound in unbelief, the virgin Mary loosed through faith (Against Heresies, 3, 22, 4).*

The Old Testament Eve was the mother of the human race in the order of nature. Mary, the mother of Jesus, is the New Eve, the new mother of the human race in the order of grace. The Old Eve was the natural mother of the human race; the New Eve, Mary, is the *supernatural* mother of the human race.

Mary is the pure Temple in which the Savior came to dwell in. In Luke 1:35 the angel of the Lord states: "The power from the Most High will overshadow (*episkiazein*) you." The phrase "to overshadow" is the same one used to describe how the cloud of God's glory came to overshadow the Ark of the Covenant (Ex. 40:35; Num. 9:18, 22).

This is not coincidental. Luke was making allusions to the Ark of the Covenant which contained the very presence of God (cf. Ex. 40). The Ark is the most holy object in all the Old Testament! The Ark was to be made "perfect in every detail" to allow that which is perfect to "fill it" (Ex. 25; 40:5). Not only did the Ark have to be perfect, it had to be kept free from all impurity and profanation (In 2 Samuel 6:6-7 Uzzah was struck dead for simply touching the Ark). For Luke, and thus for us, Mary was the pure Ark, the pure Temple, that held the divine presence, the Son of God, Jesus. And what God dwells in is pure and perfect. God does not co-exist with impurity or imperfection.

When comparing the Greek and Hebrew imagery used for the Ark of the Covenant (Ex. 25:20; 40:35; Num. 9:18, 22) and the scene of the annunciation (1 Chr. 28:18; Lk. 1:35f), one cannot but see—when read in their original languages—the powerful and unquestionable parallel. Compare Exodus 40:34-35, Numbers 9:15 with Luke 1:35; compare 2 Samuel 6:11 with Luke 1:26, 40; compare 2 Samuel 6:9 with Luke 1:43; compare 2 Samuel 6:14-16 with Luke 1:44. Can anyone doubt that Mary is the New Ark of the Covenant?

The Ark was to be made perfect for God to dwell within. Mary was created "full of grace," "without sin" so that Emmanuel, the "God who is with us," Jesus, could dwell within her.

The Ark carried the *written* Word of God; Mary carried the *living* Word of God. Ambrose writing around the year 390 said of Mary:

> *The Ark contained the Ark of the Tables of the Law; Mary contained in her womb the heir of the Testament. The Ark bore the Law; Mary bore the Gospel. The Ark made the voice of God heard; Mary gave us the very Word of God. The Ark shown forth with the purest of gold; Mary shown forth both inwardly and outwardly with the splendor of virginity. The gold which adorned the Ark came from the interior of the earth; the gold with which Mary shone forth came from the mines of heaven.*

Mary is spouse of the Holy Spirit. How can the Holy Spirit dwell in that which is sinful? Mary gave Jesus his body. This body could not carry on original sin nor concupiscence. Therefore, Mary had to be without original sin or personal sin.

Now one may argue by saying: "How can Mary have been saved prior to the crucifixion and how is it that she had no "original sin" or personal sin on her soul?"

Is Abraham in heaven? Is Isaac in heaven? What about Moses, Isaiah, Hosea, Amos, Joel, Micah, Jonah, Nahum, Ezekiel, Elijah, Elisha, Jeremiah, Habakkuk, Obadiah, Zephaniah, Haggai, Zechariah, Malachi, or Daniel? Are they in heaven? Of course they are, yet they lived before the Incarnation of the Son of God and lived before the crucifixion (Lk. 16:22; 1 Pet. 3:18f).

We must remember that Christ's salvific event, his dying for our sins on the cross, was not limited to one time period. Jesus' salvific event engulfed all of history. It engulfed that which is beyond the limits of space and time. Thus Mary being preserved from the stain of "original sin" and personal sin is not so hard to grasp in this context.

Let us not forget, Mary needed a Savior: "My spirit rejoices in God my savior" (Lk. 1:47). The Catholic Church has never denied this.

Now one may argue from Romans 3:23 that since "all have sinned" Mary must have sinned in her lifetime. But this begs the question: How would you apply Romans 3:23 to infants? Infants who are below the age of reason cannot sin, for in order to sin one must have reason and free will. To sin one must know what one is doing and have the freedom to do it. An infant has no idea of what he or she is doing and consequently cannot commit any personal sin. Romans 3:23 is a reference not to one particular individual or individuals but to the mass of humanity. Furthermore, Jesus obviously had no sin (Heb. 4:16). Paul is using "all have sinned" in the collective sense, not the distributive sense.

In Genesis 1:2 and following we are reminded that from the *immaculately* created cosmos God created Adam (Evil and chaos entering the world only after the fall, Gen. 3.). In Romans 5:14 and 1 Corinthians 15:22 we are reminded that Jesus is the second Adam. If the first Adam was created from pristine organic materials, what would the second Adam be created from? Obviously an immaculate, pristine Mother!

Brothers and sisters of the Lord?

I will pour out on the house of David and on the inhabitants of Jerusalem a spirit of grace and petition; and they shall look on him whom they have thrust through, and they shall mourn for him as one mourns for an only son and they shall grieve over him as one grieves over a firstborn.

> The Prophet Zechariah (cf. 12:10-11)

The prophet Zechariah reminds us that the Messiah, the Christ, the Savior, would be an "only child" and that he would have the privileges of the "firstborn." If Jesus had blood brothers and/or sisters then, according to this prophecy of Zechariah regarding the Messiah, the Christ, Jesus could not be the Messiah!

Mary remained a virgin throughout her life. She had no other child than Jesus.

Catholics believe in the perpetual virginity of Mary. The title "ever-virgin" has been a title for Mary from antiquity. If she had given birth to anyone else other than Jesus, that ancient title would have ceased to exist.

Even the pillars of all modern mainline Protestant denominations affirm Mary's perpetual virginity. Martin Luther wrote:

It is an article of faith that Mary is Mother of the Lord and still a virgin.... Christ we believe, came forth from a womb left perfectly intact (Works of Luther, 6, 510).

Ulrich Zwingli wrote:

I firmly believe that Mary, according to the words of the gospel, as a pure Virgin brought forth for us the Son of God and in childbirth and after childbirth forever remained a pure, intact Virgin (Zwingli Opera, v. 1, 424).

What ever happened to their Protestant offspring?

In the Scriptures and in history we never find the appellation "Mary's children." If Mary would have had children, the title "Mary's children" would certainly have been found somewhere in history. When Jesus was found in the temple at the age of twelve by Mary and Joseph, the context of the scene makes it quite clear that Jesus was Mary's only child (Lk. 2:41-51). In Mark 6:3 we are reminded that Jesus is the "son of Mary" and not "a son of Mary."

So how do we respond to quotes such as those found in Matthew 12:46, Mark 3:31-35; 6:3, John 7:5, Acts 1:14, in 1 Corinthians 9:5 and Galatians 1:19. And how do we respond to the word for brothers as *adelphoi* which means "from the womb."

First, our English word for "brother" comes from "from the same parents." Yet we use the word brother or brothers more broadly (i.e., "brothers in arms").

Secondly, when interpreting the Scriptures we need to understand the term "brother" in the same way the people during Jesus' time understood the term. When we do this, the use of the term "brother" becomes clarified. In the Hebrew and Aramaic language of Jesus' time, there was no word for cousin, uncle, close relative, and so forth. The word "brother" was used for all such appellations. It is for this reason that the word "brother" in the New Testament is used over 105 times and the word "brothers" is used more than 220 times.

In the Old Testament, or what we refer to as the Hebrew Scriptures, brothers and sisters are often meant to refer to close relations. Brothers and sisters in Semitic usage can refer to nephews, nieces, cousins, half-brothers, half-sisters, etc. (cf. Gen. 13:8; 14:14-16; 29:15, Lev. 10:4, etc.). For example, when we look at the original Hebrew texts and even the better English translations we find the following: Lot is described as Abraham's brother, yet Lot is the son of Aran (cf. Gen. 14:14). Lot was Abraham's nephew. Jacob is called the brother of Laban, yet Laban is his uncle (Gen. 29:15). When we look to Deuteronomy 23:7-8 and Jeremiah 34:9 we notice the appellation brothers is used in terms of a person who shares the same culture or national background. When we look to 2 Samuel 1:26 and 1 Kings 9:13 we notice that brother is used in terms of a friend. When we look at Amos 1:9 we see that brother is used in terms of an ally.

The New Testament makes it quite clear that this is the appropriate understanding of brothers and sisters in reference to Mary. For example, James and Joseph are called "brothers of Jesus," yet in examining the Bible we see that this is impossible, for James and Joseph "are sons of another Mary, a disciple of Christ," whom Matthew significantly calls "the other Mary," Mary the wife of Clopas (Jn. 19:25) (Mt. 13:55; 28:1; cf. Mt. 27:56). In the Acts of the Apostles, Peter in addressing the "one hundred and twenty brothers" [adelphon] (Acts 1:15f), was certainly not addressing one hundred and twenty blood-brothers! In Acts 22:7 fellow Christians are called "brothers," "adelphon," and the Jewish leaders are called "brothers," "adelphon." The Greek word adelphoi has a broad meaning as does the English understanding of the word. In fact, in the ancient world it had an ever broader meaning!

Once again, nowhere in the New Testament is there the appellation "sons of Mary." Furthermore, when we hear that the "brothers" advised and reprimanded Jesus (i.e., Jn. 7:3-4; Mk. 3:21), the idea of blood brothers of Jesus becomes even more absurd since younger brothers in the Jewish culture did not admonish older brothers!

It is not an unusual or an odd practice to call people who are not related to us as brothers or sisters. Even today in many Protestant denominations people like to refer to themselves as brothers and sisters in the faith. Yet they are not real brothers or real sisters. They are close friends within the Body of Christ. As Jesus himself mentions we are "all brothers" (Mt. 23:8).

Other examples of the use of brothers in a non-familial sense can be seen in the following: In Romans 14:10, 21, we read, "Why then do you judge your brother?" "It is good not to...do anything that causes your brother to stumble." In 1 Corinthians 5:11 we read, "I now write to you not to associate with anyone named a brother, if he is immoral, greedy...." In 2 Corinthians 8:18 we read, "We have sent to you the brother who is praised in all the churches for his preaching of the gospel." In 1 Thessalonians 4:6, we read, "Do not take advantage or exploit a brother." In 1 John 3:17 we read, "If someone who has worldly means sees a brother in need and refuses him compassion, how can the love of God remain in him?" In 1 John 4:20 we read, "If anyone says, 'I love God,' but hates his brother, he is a liar." And on and on the pattern goes. In fact, Paul makes use of the appellation "brothers" in 97 scriptural verses.

Another issue that needs to be dealt with is the issue of the phrase that we find in Luke 2:7 where Jesus is mentioned as the "firstborn." In the English language firstborn implies the birth of other children; however in the time of Jesus "firstborn" had no such implication. For example, in the Old Testament Psalm 89:27 David is referred to as the "firstborn," yet he is the eighth son of Jesse (1 Sam. 16). In the Old Testament book of Genesis (43:33), we read about Joseph as being referred to as the "firstborn." Yet this cannot be understood in the modern sense of firstborn since Joseph was one of the youngest children of the Patriarch Jacob (He was the firstborn of Rebecca but not Leah). In the book of Exodus, Moses reminds the Pharaoh, because of his obstinacy "every firstborn in the land of Egypt shall die" (Ex. 11:5). Obviously, no implication for second-born children can be inferred. (A further point is that in the ancient world the term "firstborn" often had the connotation of only-born). The firstborn son had to be redeemed within forty days (Ex. 34:20). There would be no way of knowing if other children would be born after!

The term "firstborn" is primarily a legal term, a term indicating rights and privileges (i.e., Gn. 27; Ex. 13:2; Nb. 3:12-13; 18:15-16; Dt. 21:15-17). For example, the term "firstborn" often referred to a child that was responsible for opening the womb of a woman, without any further implication (Ex. 13:2; Nb. 3:12). Sometimes it referred to someone as being special, as being sanctified (Ex. 34:20). According to the Law of Moses, all Jewish firstborn children were to be presented in the Temple and offered to God in thanksgiving (cf. Lk. 2:22f). This very fact did not mean that the parents of this firstborn child, presented in the Temple, were going to have a second-born child or any other children! Another account that explains the term "firstborn" comes from within the context of the whole of the infancy narrative (Lk. 1:5 to 2:52). Firstborn within this context is a reference to Jesus as the "firstborn of God" (cf. Col. 1:15, 18; Heb. 1:6; Rev. 1:5). Jesus is the "firstborn" of all creation (Col. 1:15).

Some like to refer to Matthew 1:25 where the phrase used is "until she bore a son." They imply that the word "until" implies children after the birth of Jesus. Again, we must look at the way this word was understood in the time of Jesus. No implication can be made regarding marital relations by the use of the word. It was a common phrase of the period which had no further implications regarding further births.

"Until" in ancient Greek is a compound word *heos-hou*. This compound word implies no implication of further events. For example, in Luke 1:80 John the Baptist was called by God to remain in the desert "until the day of [the Messiah's] manifestation to Israel." Yet John remained in the desert after the Savior's manifestation to Israel and even after Jesus himself began baptizing in the Jordan—only with John's capture by Herod's men, did he cease to be in the desert. In Acts 25:21 Paul was to remain imprisoned "until" (*heos-hou*) he was sent up to Caesar. Yet the Acts of the Apostles (cf. 28:20) show Paul remaining in custody even after his meeting with Caesar. In the Septuagint version of Isaiah 46:4 God says: "I am until you grow old." Did God cease to exist when Isaiah grew old? NO! When God the Father spoke to his Son saying, "Sit on my right hand until I make your enemies your footstool," he certainly was not implying that the Son would no longer sit on his right hand once his enemies were restrained. In the Septuagint version of Psalm 111:8, the version the early Christians used, we read of a man whose heart is steadfast and who "shall not be afraid until (*hous-hou*) he looks down upon his foes." Obviously this man will not suddenly become afraid after conquering his foes. If anything, he would have been afraid before gaining dominance over his enemies. In 2 Peter 1:19 Peter reminds the faithful to remain dutiful to God until (*hous-hou*) the "day dawns and the morning star rises in your hearts." If "until" had further implications, then Peter would be saying that once the morning star rises we can forget about God's word.

Another word for "until" used in the Scriptures is *heos*. In Mathew 28:20 Jesus reminds the faithful that he will be with them until (*heos*) the end of the world. Does that mean that at the end of the world Jesus, God, will disappear? Obviously not! In 1 Corinthians 15:25 Paul states that Christ will reign as king until (*heos*) he has positioned his enemies under his feet. Will he cease to reign as king once he has triumphed over his foes? In 1 Timothy 4:13 we hear Paul exhort Timothy to preach and teach the faith until (*heos*) Paul's arrival. Does that mean that once Paul arrived Timothy would never again preach the faith? Clearly, Timothy continued to preach the faith even after Paul's arrival.

The word "until" has no further implications in the Greek of Jesus' time. The purpose of verse 25 was to emphasize that Joseph was not in any way responsible for the birth of Jesus.

The most powerful argument however for the perpetual virginity of the Blessed Mother is found at the foot of the cross.

When Jesus saw his mother and the disciple there whom he loved, he said to his mother, "Woman, behold your son." Then he said to the disciple [John], "Behold, your mother." And from that hour the disciple took her into his home (Jn. 19: 26-27).

It makes absolutely no sense for Jesus to give his mother over to the apostle John if a brother or brothers or sister or sisters were around. Wouldn't you entrust your mother to a brother or sister? And would a mother abandon her own children so as to become the mother of another? As Athanasius of Alexandria (ca. 295) explained:

If Mary had had other children, the Savior would not have ignored them and entrusted his Mother to someone else; nor would she have become someone else's mother. She would not have abandoned her own to live with others, knowing well that it ill becomes a woman to abandon her husband and her children. But since she was a virgin, and was his Mother, Jesus gave her as a mother to his disciple, even though she was not really John's mother, because of his great purity of understanding and because of her untouched virginity (De virginitate, in Le Museon 42: 243-44).

In Matthew 15 Jesus condemns the Pharisees for the "Korban rule," a rule that allowed children to avoid taking care of their parents. Jesus would not have brothers or sisters who would ignore the taking care of Mary.

Hence, Mary is the "ever-virgin" as the Church has from the beginning of time always called her. It is significant that this belief was so universally held that it only began to be questioned in the year 380 when a man by the name of Helvidius saw "brothers and sisters" as being real brothers and sisters. Jerome, who translated the Bible from the original languages into Latin, felt that Helvidius' interpretation was so ridiculous that it was not worthy of a single comment.

There is no child of Mary except Jesus...

Origen (ca. 250)
Commentary of John I:4; PG 14, 32; GCS 10, 8-9.

Why was Mary assumed into heaven?

Even though your most holy and blessed soul was separated from your happy and immaculate body, according to the usual course of nature, and even though it was carried to a proper burial place, nevertheless it did not remain under the dominion of death, nor was it destroyed by corruption. Indeed, just as her virginity remained intact when she gave birth, so her body, even after death, was preserved from decay and transferred to a better and more divine dwelling place. There it is no longer subject to death but abides for all ages.

John Damascene (ca. 645)
Homily 1 on Dormition 10: PG 96, 716 A-B

Jesus was "full of grace" (Jn. 1:14) and "without sin" (Heb. 4:15). Jesus ascended body and soul into heaven (Lk. 24:50-53). Mary, being the perfect disciple, the perfect imitator of her son, the perfect model of the Church, the one who knew her son more than any other creation of God, would be granted the gift of imitating her Savior, her son, by being "full of grace" and without sin. And at the end of her earthly journey, she would imitate her Savior, her son, by being assumed by Him into heaven body and soul.

The fact that one could be raised without decay should not be troubling to a Christian. In Matthew 27:52 we are reminded that many saints who had fallen asleep were raised—without decay taking place to their bodies—at the Crucifixion. In 1 Thessalonians 4:17 we are reminded that many will be caught up to meet the Lord in the air body and soul.

The fact that Mary was assumed into heaven should not be a shocking idea for the Christian. After all, in Genesis 5:24 and Hebrews 11:5 we read how Enoch was "taken up" to God. And in 2 Kings 2:1, 11, we are told how Elijah was taken up to heaven in a whirlwind. If Enoch and Elijah are taken up to God, why would we have trouble believing that the Mother of Jesus, the Mother of the Savior would be taken up, assumed, into heaven.

The teaching of Mary's assumption into heaven is the belief that Mary after the course of her earthly life was assumed body and soul into heaven.

In the ancient Byzantine Liturgy of the Catholic Church we hear the following liturgical expression by the Eastern Fathers on the Feast of the Dormition:

> *In giving birth you kept your virginity; in your Dormition you did not leave the world, O Mother of God, but were joined to the source of Life. You conceived the living God and, by your prayers, will deliver our souls from death (CCC 966).*

Original sin and personal sin are what prevent a person from entering into heaven. But because of the merits of Jesus, Mary was "full of grace" and therefore without "original sin" or personal sin. Heaven was open to her, and because of her special place in the life of Jesus, she was assumed into heaven.

Mary is spouse of the Holy Spirit. How can the Holy Spirit dwell in that which is sinful? Mary gave Jesus his body and cooperated with God in giving Jesus his soul. This body and soul could not carry on "original sin" nor concupiscence. Therefore, Mary had to be without "original sin" or personal sin.

When we look to the historical evidence of those who were close to Jesus we notice that their bones are venerated and held in places of honor in churches throughout the world. Yet no mention has ever been made about the bones of Mary and no mention has ever been made about the veneration of her bones anywhere.

Mary knew no decay, for she was free from original sin and concupiscence. As Psalm 16:10 reminds us: "[the beloved will not] know decay." The beloved blessed Mary knew no decay. She was assumed body and soul into heaven.

Just as the Ark of the Covenant was to remain intact, the New Ark of the Covenant, Mary, was to remain intact, Body and Soul!

The Bible, logic, and history point to Mary's assumption.

It is interesting to note that August 15 has always been reserved by Lutherans and Anglicans as a day for Mary. In recent years, Anglicans have allowed their followers to believe in the Assumption; they concluded that this belief was in perfect conformity with the Scriptures. Likewise, some branches of Lutheranism allow for the belief in the Assumption—as a matter for personal devotion.

Mary, the "woman" (cf. Jn. 2:4; 19:26)

Mary is the "woman" of Genesis 3:15, who with her son, will crush the serpent's head; she is the "woman" at the foot of the cross at Calvary, the "skull-place," where the serpent's head is **crushed**. Mary is the "woman" who obeys God as opposed to the "woman" who disobeyed God in the Garden. Mary is the "woman" who wages war against the **dragon**, the serpent, the devil (cf. Rev. 12).

How much should we love Mary?

Pope Benedict XVI, in his ecumenical discussions, has found that more and more Protestants are turning back to Mary, for they recognize that a Christianity without Mary is a Christianity that is lacking, lacking the feminine dimension of human life.

Blessed Mother Teresa of Calcutta was about to give a talk at a conference when a young woman rushed to her side and mentioned that one of the hot topics at the conference dealt with the issue of Mary and how much we should love her. Some were arguing that we were showing too much love for her, others argued we were not showing enough love for her. The debate went back and forth with no resolution in sight. Mother Teresa looked at the messenger and said that she need not worry. She would take care of the dilemma. Mother Teresa proceeded to the stage and began to address the impasse in a very simple fashion. She said: "You want to know how much to love Mary? I'll tell you. Love her *no more* or *no less* than Jesus loved her." Wow! How can we poor creatures ever equal the love of Jesus? We can never love Mary too much, for she always leads us to her Son.

Let us not forget that Jesus was obedient to Mary at the wedding feast of Cana and at the temple. God was obedient to Mary.

Let us never forget Mary's special place in history. Remember to honor Mary is to honor her son, Jesus, for it is Jesus who gave Mary all her privileges. To deny Mary's privileges is to deny Jesus' will and work on behalf of Mary! In honoring Mary we are being obedient to God's will of honoring her; Mary's privileges were given by God and not man. Let us be obedient to the Scriptures which remind us that "all generations will call me blessed" (cf. Luke 1:48).

We as Christians are members of the Body of Christ, the Church (1 Cor. 12). Mary is the Mother of the Head of the Body, Jesus. So Mary is our mother too! She is the Mother of the Church, Christ's Body!

VI
END-TIME ISSUES

Hell, a reality

Do not doubt, the evil will depart this world and enter into the unquenchable fires of hell.

Ignatius of Antioch (ca. 107)
Disciple of the apostle John

Many groups refuse to believe in the reality of hell. They argue that an all-loving God would never condemn anyone to hell.

This might seem to be culturally and politically comforting but the reality of hell is something that one cannot run away from.

To argue that because God is all-loving he would never condemn anyone to hell is a misunderstanding of the nature of love. There is no such thing as love without justice, and justice demands that those who accept God's grace are to be rewarded for that acceptance, and those who blatantly reject God's grace are to receive their desire, life without God in hell.

Those who deny the existence of hell deny the preciousness of the gift of grace, of free will, and the reality of authentic love, which cannot be divorced from justice.

God loves us so much that he respects our free will decision to choose, in response to grace, eternal life with him or eternal life without him.

The reality of hell as a place of endless suffering is attested to throughout the Bible (Is. 66:24; Mt. 8:12; 25:41; Lk. 16:25; 13:28; 2 Thess. 1:9; Rev. 14:10; 18:7).

We cannot be, as the late John Cardinal O'Connor argued, "Cafeteria Catholics" or "Cafeteria Christians." We cannot pick and choose what we like or don't like. Catholicism is an all or nothing faith. Either we accept all of its infallible teachings or we might as well join a superficial version of Christianity where one can believe in anything.

The denial of the existence of hell is a sin, the sin of presumption. The sin of presumption distorts the relationship between God's mercy and justice.

Where do we get purgatory from?

Some like to argue that since the word "purgatory" is not in the Bible, it does not exist. The word "Bible," "Incarnation," and "Trinity" are nowhere to be found in the Bible, yet they are part of Christianity's experience of reality. The word may not be present but the reality certainly is!

On the next day [after the battle with Gorgias]...Judas [Maccabee] and his men went to take up the bodies of the fallen and to bring them back to lie with their kindred in the sepulchers of their ancestors. Then under the tunic of each one of the dead they found sacred tokens of the idols of Jamnia, which the law forbids the Jews to wear. And it became clear to all that this was the reason these men had fallen. So they all blessed the ways of the Lord, the righteous judge, who reveals the things that are hidden; and they turned to supplication, praying that the sin that had been committed might be wholly blotted out. The noble Judas exhorted the people to keep themselves free from sin, for they had seen with their own eyes what had happened as a result of the sin of those who had fallen. He also took up a collection, man by man, to the amount of two thousand drachmas of silver, and sent it to Jerusalem to provide for a sin offering. In doing this he acted very well and honorably, taking account of the resurrection. For if he were not expecting that those who had fallen would rise again, it would have been superfluous and foolish to pray for the dead. But if he was looking to the splendid reward that is laid up for those who fall asleep in godliness, it was a holy and pious thought. Therefore, he made atonement for the dead, so that they might be delivered from their sin (2 Macc. 12:39-46, NRSV).

The Jewish feast of Chanukah is a commemoration of this battle and the cleansing of the temple that followed this event. This is a passage that would have been in the hearts and minds of all the apostles.

In this passage we see that Judas Maccabee finds that the dead on the battlefield have sinned by wearing amulets associated with a false god. Now Judas does something critically important: he prays for the forgiveness of their sins; he takes a collection for an expiatory sacrifice, and seeks atonement so that those who died in battle in sin might be freed from sin.

This passage is the most powerful proof for purgatory in all of the Scriptures. If one dies, one either goes to heaven, purgatory, or hell. If one goes to heaven, one has no need for one's sins to be blotted out, since one is enjoying eternal paradise. If one is in hell, then all the prayers in the world cannot release one from hell since hell is eternal (Mt. 25:41; 2 Thess. 1:9). Hence, if sin can be blotted out after death, then there needs to be a place for purification; that place is called purgatory.

The Bible makes reference quite often to a "cleansing fire" (i.e., cf. 1 Pet. 1:7; Wis. 3:1-6). We would call this cleansing fire purgatory, where God's fiery love "burns" away the soul's impurities, where one is "saved, but only through fire" (1 Cor. 3:1-16). The Bible also testifies to the reality of paying debts, as in the case of the Judge who reminds us that we "will not be released until [we] have paid the last penny" (Mt. 5:21-26; also 18:21-35; Lk. 12:58;

16:19-31; 1 Pet. 3:19; 4:6). In other words, we will not be released until every sin, every word, is accounted for (Mt. 12:36).

Gregory the Great (ca. 540-604) in reflecting on Matthew 12:31-32 explains with great insight the following in regards to purgatory:

> *As for certain lesser faults, we must believe that, before the Final Judgment, there is a purifying fire. He who is truth says that whoever utters blasphemy against the Holy Spirit will be pardoned neither in this age nor in the age to come. From this sentence, we understand that certain offenses can be forgiven in this age, and certain others in the age to come (Dial. 4:39: PL 77, 396).*

Historically, prayers for the dead have always been part of Hebrew and Christian tradition. In the Jewish Orthodox culture, prayers for the dead were common. At the time of Jesus, prayers for the dead in the Jewish faith were said in temples and synagogues on feasts such as Passover, Booths, and Weeks. Jews to this very day still utter the "Mourner's Kaddesh" after the death of a person for the purification of his or her soul.

Graffiti in the catacombs of Rome from the first three centuries of Christianity, when the Church was under persecution, attests to this common practice. In the first century catacombs we read: "Sweet Faustina, may you live in God." "Peter and Paul, pray for Victor."

Basil the Great in 370 illustrates that the very nature of the Christian life, as a struggle and battle, makes the reality of purgatory an absolute necessity:

> *A man who is under sentence of death, knowing that there is One who saves, One who delivers, says: 'In You I have hoped, save me' from my inability 'and deliver me' from captivity. I think that the noble athletes of God, who have wrestled all their lives with the invisible enemies, after they have escaped all of their persecutions and have come to the end of life, are examined by the prince of this world; and if they are found to have any wounds from their wrestling, any stains or effects of sin, they are detained. If, however, they are found unwounded and without stain, they are, as unconquered, brought by Christ into their rest (Ps. 7, n. 2).*

Augustine of Hippo in 387 records in his masterpiece *Confessions* the words of his mother, Monica: "All I ask of you is that wherever you may be you will remember me at the altar of the Lord." In other words, prayers were to be said for her, the prayer of the Mass.

In Augustine's *De fide, spe, caritate liber unus* (39, 109) we read:

> *The time which interposes between the death of a man and the*

final resurrection holds souls in hidden retreats, accordingly as each is deserving of rest or of hardship, in view of what it merited when it was living in the flesh. Nor can it be denied that the souls of the dead find relief [in purgatory] through the piety of their friends and relatives who are still alive, when the Sacrifice of the Mass is offered for them, or when alms are given in the church. But these things are of profit to those who, when they were alive, merited that they might afterwards be able to be helped by these things. For there is a certain manner of living, neither so good that there is no need of these helps after death, nor yet so wicked that these helps are of no avail after death. There is, indeed, a manner of living so good that these helps are not needed [in heaven], and again a manner so evil that these helps are of no avail [in hell], once a man has passed from this life.

It is not until the Protestant Reformation that prayers for the dead become seriously challenged.

Pure logic attests for the need of a place called purgatory. We are reminded to be "perfect as [the] heavenly Father is perfect" (Mt. 5:48). We are called to "strive for that holiness without which one cannot see God" (Heb. 12:14). In Revelation 21:27 we are told that "nothing unclean shall enter heaven." Heaven is a place of perfection where "nothing impure" can enter (Rv. 21:27). If this is so, then one who dies in sin must be purified. In Hebrews 12:33 we are reminded that the "spirit of the just are made perfect;" that is made perfect to enter into heaven. If anything impure were to enter into heaven, then heaven would no longer be a place of purity for it would have been tainted with impurity. Pure water, for example, if it is contaminated with a chemical, is no longer pure water. Likewise, heaven, if it is contaminated with imperfection, is no longer a place of perfection.

Another interesting clue to purgatory is seen in 1 Corinthians 15:29-30 where people were baptizing themselves for their dead loved ones. This very practice—albeit a wrong practice—by some within the Christian community points to early Christianity's recognition that there had to be something more than just heaven or hell. There had to be a place where the efforts of the living on earth could have an impact on those in the afterlife. We call this place purgatory.

Another fascinating example is that of Jesus preaching to the "dead" after the crucifixion (1 Pet. 3:18-20: 4:6): "For this is why the gospel was preached even to the dead, that, though condemned in the flesh in human estimation, they might live in the spirit in the estimation of God." Clearly Jesus was cleansing, purging them through his preaching to enter into heaven!

Hence from a Scriptural point of view and from a philosophical point of view, derived from the Scriptural understanding of heaven, purgatory is a reality of Christianity.

In purgatory one is assured of heaven, but is purified to enter the realm of perfection. In purgatory a man gains, as Gregory of Nyssa (ca. 379) states, "knowledge of the difference between virtue and vice, and finds that he is not able to partake of divinity until he has been purged of the filthy contagion in his soul by the purifying fire" (*Sermones*, 1).

How sad it must be for those who cannot pray for their deceased loved ones. There is no greater sense of psychological closure than to pray for one that has passed away. That is why Paul asked for mercy on the soul of his dead friend Onesiphorus in 2 Timothy 1:16-18. He knew his prayers would release him from that place of purification that has become known as purgatory. (If Onesiphorus was in heaven he would be in no need of prayer, and if he was in hell no amount of prayer could release him. Therefore, Onesiphorus was in purgatory where Paul's prayers could be effective.).

Is there such a thing as temporal punishment?

Some have made the argument that when God forgives, he forgives, and therefore there is nothing to be made be up, or accounted for, or paid off. They therefore deny the reality of purgatory and the reality of temporal punishments. The above makes it quite clear that there is such a thing as purgatory, but let us address the issue of temporal punishment separately.

Temporal punishment refers to earthly punishments that flow from sin. A sin may be forgiven, but the punishment due for that sin is either to be made up for, or accounted for, or paid off in this life or in the life to come (cf. Mt. 5:26; 12:36; 1 Cor. 3:15; Rev. 21:27). As Catherine of Genoa would say, "We either do our purgatory here on earth or in the afterlife." The Bible is replete with examples of temporal punishment. But let us look at just one:

In 2 Samuel 12:13-18 we see the consequences of David's sin of adultery with Bathsheba and David's subsequent killing of her husband Uriah. The prophet Nathan reprimands David, then David says,

> '*I have sinned against the LORD.' Nathan answered David: 'The LORD on his part has forgiven your sin: you shall not die. But since you have utterly spurned the LORD by this deed, the child born to you must surely die.' Then Nathan returned to his house. The LORD struck the child that the wife of Uriah had borne to David, and it became desperately ill. David besought God for the child. He kept a fast, retiring for the night to lie on the ground clothed in sackcloth. The elders of his house stood beside him urging him to rise from the ground; but he would not, nor would he take food with them. On the seventh day, the child died.*

Notice that even though David's sins were completely forgiven, there was still a temporal, earthly, punishment associated with the sin. Though forgiven, he was still punished.

Temporal punishment is an essential part of the Christian life for God is a loving father who does what every loving father does. He calls us to live a good and holy life. And when we are disobedient he chastises us for our own good and the good of our soul. In Wisdom 3:1-6 we see how God allows for chastisements for our very good:

> But the souls of the just are in the hand of God, and no torment shall touch them. They seemed, in the view of the foolish, to be dead; and their passing away was thought an affliction and their going forth from us, utter destruction. But they are in peace. For if before men, indeed, they be punished, yet is their hope full of immortality; Chastised a little, they shall be greatly blessed, because God tried them and found them worthy of himself.

David paid a heavy price for his sin, but he also changed as a person because of the temporal punishment that was inflicted upon him. David would become one of the Bible's greatest saints.

Is hell eternal or are the souls of the evil annihilated?

All orthodox Christians believe in a hell, but some groups such as the Jehovah's Witnesses and the Seventh Day Adventists argue that at death the soul "rests" or is "suspended in existence" until it is resurrected on the last day. Those that are good will enjoy eternity with God in heaven, while the evil will be condemned and their souls annihilated.

For the Catholic or the orthodox Christian the soul lives eternally from the moment of conception, either eternally with God or eternally without God, depending on one's response to grace.

For mainline Christians, the pains of hell are endless and the soul is immortal. In Isaiah 33:11, 14 we are reminded of the everlasting flames of hell; in Matthew 3:12 hell is a place where the damned "burn with unquenchable fire"; in Matthew 25:41 we read, "Depart, you accursed, into the eternal fire." In Matthew 25:46 we are reminded that the evil "will go off to eternal punishment." In Luke 3:16-17 the evil "will burn in unquenchable fire." In 2 Thessalonians 1:6-9 the evil will pay the penalty of "eternal ruin." In Revelation 20:10 we read: "The Devil who had led them astray was thrown into the pool of fire and sulfur, where the beast and the false prophets were. There they will be tormented day and night forever and ever." Our souls are immortal and therefore hell is everlasting for the evil.

Furthermore, for mainline Christians there is a distinction between the particular judgment (which occurs immediately with death) and the general or last judgment (which occurs at the second coming of Christ). When we die (death being when the soul and body are separated), we are immediately judged and our soul enters into heaven, purgatory, or hell (cf. Mt. 17; Lk. 16:19-31; Heb. 9:27; 10:31; Wis. 3:1f; Eccl. 8:6-8; 11:9; 12:14; Sir. 11:28f; 16:13-22; 2 Macc. 12:43-46); the soul is not "suspended" or at "rest" until the end of time. In Luke 23:42f we read: "[The thief hanging next to Jesus] said, 'Remember me when you come into your kingdom.' He replied to him, 'Amen, I say to you, today you will be with me in paradise.' Jesus assured the thief that today was his day! The thief's soul would be shortly enjoying the gift of paradise, while his body would begin to decay in the tomb.

At the end of time, at the last or general judgment, our body will be resurrected in a glorified form (Phil. 3:21) and be reunited with our soul in heaven (cf. Rv. 6:14-17; 7; 8; 9). The last judgment and the resurrection of the bodies of the dead is well attested to in many scripture passages such as Acts 24:15, "I have the same hope in God as they themselves have that there will be a resurrection of the righteous and the unrighteous." Judgment will follow. In Philippians 3:20-21 we read about the gift of the glorified body, "Our citizenship is in heaven, and from it we also await a savior, the Lord Jesus Christ. He will change our lowly body to conform with his glorified body by the power that enables him also to bring all things into subjection to himself."

For those who are still living at the time of Jesus' second coming, the parousia, they will be judged body and soul at that very moment (cf. Mt. 25:31f).

Heaven and hell are eternal. A person is born to eternal life, eternal life with God or eternal life in hell, depending on his or her response to grace.

The concept of the annihilation of souls is found in eastern religions and particularly in the Christian heresy of Gnosticism—with its idea of inferior gods. For the Gnostics, the non-elect, the hylikoi, would be annihilated.

Jehovah's Witnesses likewise have a similar belief in the annihilation of the soul. For Jehovah's Witnesses a man does not have a soul but is a soul. The death of man is the death of the soul. At the end of the world the good shall be created anew.

What the Jehovah's Witnesses fail to recognize is that there are three ways of understanding the soul in the Bible. It is often referred to as the seat of wisdom and emotions (Ps. 42:2), as a whole person (1 Peter 3:20), and as a life principle (Mt. 10:28; Rev. 6:9-10).

When humans were created they were created in the image and likeness of God (Gen. 1:26f). God is immortal and so are we! In Matthew 17:1-8 we see Jesus, Moses, and Elijah conversing. Clearly their souls were not annihilated with death! In Luke 16 we read about the resurrection of Jesus, body and soul. Undoubtedly Jesus' soul is not annihilated! In Revelation 6:9-10 we read: "...I saw underneath the altar the souls of those who had been slaughtered because of the witness they bore to the word of God. They cried in a loud voice, 'How long will it be, holy and true master, before you sit in judgment, and avenge our blood on the inhabitants of the earth.'" Clearly these people were not annihilated. Clearly their souls survived them and they are in fact immortal.

For Christians hell is permanent; there is no annihilation of souls or the ceasing of the existence of hell or the ceasing of the existence of the person (Mt. 3:12; 9:43; Rev. 20:10; Mt. 18:8; Mt. 25:46).

These groups are making the same mistake that those who were around Jesus were making. Jesus always reprimanded the Sadducees for their lack of belief in an afterlife; whereas, he affirmed the Pharisees in their belief in the afterlife.

Orthodox, Catholic Christianity (as well as Protestant Christianity) has always viewed the annihilation of souls as heretical and unchristian. Ignatius of Antioch (110), the friend of the apostle John, reminds the evil of the pain of the eternal "unquenchable fire" of hell (Ephesians 16:1). Polycarp, the friend of the apostle John, reminded the faithful to keep their eyes on Christ in order to avoid the "eternal and unquenchable fire" of hell (2:3). The list is unending regarding the eternity of hell!

Premillennialism, Postmillennialism, Amillennialism and the Rapture

Revelation 20 and 1 Thessalonians 4:15-17 are among the most fascinating passages in Scripture in that they have been interpreted in such radically different ways. The primary reason for this is the confusion over the placing of the thousand year reign, the rapture, and over the sense in which these passages were meant to be understood.

Premillennialism

Premillennialism holds that after the period of the Church there will be a time of tribulation that will be followed by Christ's Second Coming, the binding of Satan, and the resurrection of the faithful who have died in Christ. Christ and the risen faithful will reign on earth physically for a thousand years. This will be followed by another period of tribulation, albeit short, the Final Judgment, and the rapture of the faithful into heaven. The creation of a new heaven and a new earth will follow.

Postmillennialism

Postmillennialism holds to the idea that after the period of the Church, Satan will be bound, and a thousand year reign will follow, followed by the rapture into heaven of the living faithful before the period (or during the middle) of the tribulation. This will be followed by the Second Coming of Christ, the resurrection of the dead, the Final Judgment, and the creation of a new heaven and a new earth.

Amillennialism

Catholicism rejects both Premillennialism and Postmillennialism. It believes in what is called Amillennialism. It holds that Revelation 20 is a symbolic passage and that the thousand year reign is a symbolic term for the period from Christ's salvific act to the time of Christ's Second Coming. Christ's Second Coming will be preceded by a short tribulation period. Jesus' return will be followed by the resurrection of the dead (Acts 24:15), the Final Judgment (Mt. 25:31, 32, 46; Jn. 5:28-29; 12:49) and the creation of a new heaven and a new earth (Rom. 8:19-23; Eph. 1:10; 2 Pet. 3:13; Rev. 21:1-2, 4-5, 9-27) How the transformation of a new heaven and a new earth will take place and how it will look like or when the Second Coming will occur is part of the mystery of our faith.

The Catholic Understanding of the Rapture

And what is the Catholic understanding of the rapture as found in 1 Thessalonians 4:15-17? At the second coming of Christ, the dead will be resurrected, the Final Judgment will take place, and those faithful who are still alive when Christ returns—and after the Final Judgment—will go up with the resurrected faithful to meet and be with Christ forever.

The resurrection of the dead at the end of time is a reference to the resurrection of the bodies of the righteous and unrighteous (cf. Acts 24:15). The bodies of the righteous will be reunited in a glorified form (Phil. 3:21) to their souls in heaven; the bodies of the unrighteous will be reunited to their souls in hell. Those still living in body and soul at the end of time will be judged at that time (the Final Judgment) and then follow Christ in body (in a glorified form) and soul into heaven or go body and soul into hell.

- Millenarianism, the idea of an earthly thousand year reign by Christ, has always been rejected in the Church. The only early writer to take the idea of an earthly thousand year reign seriously was Papias, and he was ridiculed for taking a position so contrary to the Church's traditional interpretation.

What about the 144,000?

Basing their beliefs on Revelation 7:1-8 and Revelation 14:1-5, Jehovah's Witnesses and other pseudo-Christian groups and fundamentalists argue that only 144,000 people will be saved.

The 144,000, for virtually all mainline Christians, is a symbolic number, for if it were not, Jehovah's Witnesses, fundamentalists, and most Christians would surely not be among the 144,000, for the 144,000 are made up of only male virgins according to Revelation 14:4. And since these religious groups are not made up of entirely male virgins, they will not be counted among the 144,000. Furthermore, in the first 300 years of Christianity alone, more than one million Catholic Christians were martyred for their faith in Jesus Christ. Are they not in heaven? Of course they are! It would be nonsensical to think otherwise!

For Jehovah's Witnesses there are two groups of people: the 144,000 "anointed," and the "other sheep." The "anointed" go to a heavenly paradise and the "other sheep" go to an earthly paradise.

This is contradicted by the Bible. How do the Jehovah's Witnesses explain 2 Kings 2:11 or Hebrews 11:5 or Luke 13:28? According to the Jehovah's Witnesses the Old Testament people could not reach the heavenly paradise—they could only live in the earthly paradise for they are part of the "other sheep." Yet these passages from the Old Testament say that Enoch and Elijah were both taken up to heaven! And in Luke 13:28 Jesus reminds us that Abraham, Isaac and Jacob are in heaven!

Furthermore, if you read a little further on you see that there is a "great multitude which **no one could count**" standing before the throne of the Lamb in heaven (cf. Rev. 7:9). In 1 John 3:2 we read that all the faithful will see God in the beatific vision, "see him as he is." In Philippians 3:20 we are told that our citizenship is in heaven. In Matthew 5:11-12 we are enlightened to the fact that the persecuted will receive a great reward in heaven—just like the one million who died painful, barbaric, and tortuous deaths for the faith in the first 300 years of Christianity!

The real and traditional interpretation for the number 144,000 is as follows: 144,000 comes from the square root of 144, which is twelve, the number of Israel's tribes, multiplied by a thousand, the symbolic number for the New Israel that embraces all peoples, races, tongues, and nations (cf. Rev. 14:1-5; Gal. 6:16; Jas. 1:1).

The 144,000 is a symbolic number that denotes that those who remain loyal to Christ, even amidst trials and tribulations, even amidst the most severe of persecutions, will be saved (cf. Phil. 3:17-21; Rom. 8:16-17; 2 Pet. 1:1-10; 1 Jn. 3:1-3). Remember, in the first 300 hundred years of Christianity, over one million people were martyred, died for the faith of Christianity!

Celestial marriages?

The idea of celestial marriages is an invention of the 19th century founder of Mormonism, Joseph Smith. It was completely unheard of till then. Basing themselves on a faulty interpretation of 1 Corinthians 15:40-42, Mormons believe that there is a *Telestial* kingdom where the non-believers go to, a *Terrestrial* kingdom for non-Mormons or lukewarm Mormons, and a Celestial kingdom for the righteous Mormons. The wicked, after a temporary period in hell, go into the *Telestial* kingdom.

To become part of the Celestial kingdom you must be a baptized Mormon, follow Mormon teaching, and be married in a Mormon Temple. Interesting to say the least: this would exclude Jesus and the apostles John and Paul!

Those whose marriages have been "sealed" in a Mormon temple will be part of the Celestial kingdom and have an eternal marriage where they will beget "spiritual children" in the afterlife. They will also become gods of their own worlds.

This belief system is contrary to the Scripture's understanding of the world and the human person. Jesus reminds us that there is no marriage or sex or child-bearing in heaven. As Jesus says in Matthew 22:29: "You are misled because you do not know the scriptures or the power of God. At the resurrection they neither marry nor are given in marriage but are like the angels in heaven." Furthermore, Jesus praises celibacy in Matthew 19:12 and the apostle Paul recommends celibacy for full-time ministers in 1 Corinthians 7:32-35. And in 1 Timothy 5:9-12 the older widows are encouraged to take a pledge of celibacy and become enrolled in the order of religious or widows. Again, if we were to follow the logic of Mormonism there would be no Jesus in heaven since he was celibate!

But what about 1 Corinthians 15:40-42? For Catholics and mainline Christians 1 Corinthians 15:40-42 is interpreted in the following manner: In Jewish tradition the world was understood to have heavenly bodies like the stars, the sun, the moon, and earthly bodies, the things that dwell on the earth (cf. 1 Enoch 18:13-16; 21:3-6; Philo, *De plant.* 12). Paul uses this understanding of the world to explain the qualitative differences of the human body before and after death, after the resurrection. Before death, a body is animated by a lower, natural life principle—*psyche*—and endowed with the properties of corruptibility and weakness. After death, the body is animated by a higher life principle (*pneuma*) and thus endowed with incorruptibility, glory, power, etc.—the qualities of God himself. In other words, after the resurrection of our bodies at the end of time, our bodies will be glorified. As St. Paul explains in his letter to the Philippians (3:20-21): "Our citizenship is in heaven, and from it we also await a savior, the Lord Jesus Christ. **He will change our lowly body to conform with his glorified body...**"

There is no foundation for Mormon beliefs in the Bible or in Christian history!

What are we to think of Bible prophecy?

Bible Prophecy has become the rage among many Pentecostals. Many books have come out in recent years predicting the end-times. Many television programs find popular support regarding end-time issues. This is not new, for every generation has made predictions about the end of the world and the second coming of Christ. Many claim that all we have to do is look at the signs of the times and we will find indications regarding the end. They often quote Mark 13:39 in their favor.

Yet history has shown us what Christ has taught us, that only the Father in heaven knows the hour or time, and that the end will come like a thief in the night when we least expect it (Mt. 24:36; 2 Pet. 3:10; Rev. 3:3).

Let us not bother ourselves with predicting the future from Bible quotations. Rather let us be content with living every moment in the present with our Lord and Savior, and let us encourage others to do the same. Instilling fear, which is often the subconscious or conscious motive behind Bible prophecy, is not a proper motive. The great Catholic mystics of the Church have always shown us by their lives that as one grows in holiness, one is less concerned with going to hell as much as with not loving enough.

Let us love, according to God's understanding of love, according to authentic love.

VII
SALVATION

Do Catholics have a different understanding of the human person than Protestants?

Yes. Catholics and Protestants—at least the original Protestants such as Luther and Calvin—have a different understanding of "original sin," and because of this difference of understanding, we have a differing view of the human person.

Original sin was the sin committed by Adam and Eve in the Garden of Eden. At the heart of "original sin" is disobedience to God and the lack of trust in God: "[S]educed by the devil, [man] wanted to be "like God," but "without God, before God, and not in accordance with God" (cf. Gen. 3:5; CCC 398).

Original sin led to the distortion of all of creation.

Because "original sin" was passed down to all generations (Ps. 51:7), all human nature was affected with the loss of original holiness and justice. The gates of heaven were closed.

By the merits of Christ, his Passion, eternal life was restored to us.

In baptism "original sin" is washed away, although the consequences of "original sin," such as suffering, illness, decay, etc., remain. The human person's nature was not completely destroyed by "original sin"; it was simply wounded (cf. Mk. 1:4; 16:16; Jn. 3:5; 1:33; Acts 2:38; 8:12f; 16:33; Rom. 6:3-6; Gal. 3:27; 1 Cor. 6:11; Eph. 5:26; Col. 2:12-14; Lk. 3:3; Heb. 10:22).

This last point is the crucial difference between Catholic theology and Protestant theology. For the Protestant, "original sin" did not simply wound human nature; they argue that it radically destroyed and perverted human nature. Thus one lost free will and the ability to count on one's reasoning abilities.

This is why Protestantism would resort to a theology that emphasized faith alone, the Bible alone, and the predestination of the elect, or a variation on this theme, the absolute assurance of salvation.

The Catholic position is much more optimistic. Human nature, as was mentioned, was wounded by "original sin," but not destroyed or perverted. It "is wounded in the natural powers proper to it; subject to ignorance, suffering, and the dominion of death: and inclined to sin, an inclination to evil called concupiscence" (CCC 405).

Since human nature is only wounded, one has not lost the gift of free will, despite an inclination to evil, and one has not lost the gift of reasoning, despite its being subject to ignorance. Thus, one can use one's natural reasoning abilities to understand one's faith and the mystery of creation. One can even reason to a knowledge of a divine reality (cf. Rom. 1:19-20; Wis. 13:1-9). One is free to love and respond to God and one is free in seeking to uncover the hidden mysteries of God. This is a much more optimistic understanding of human nature.

The following quotes should be helpful in understanding the Catholic understanding of the human person in reference to the Fall: Genesis 3:5-17, 19; 2:17; 4:3-15; 6:5, 12; 8:21; Job 14:4; 15:14; Psalm 51:7; Sirach 8:5; Romans 1:18-32; 3:9, 23; 5:12, 17, 18; 6:23; 8:6-8, 11, 13, 17, 21; Galatians 5:17; Ephesians 2:3.

"Proofs" for existence of God

Because of a different understanding of human nature, the Protestant understanding of the human person is very different than the Catholic understanding of the human person. Protestants and Catholics have different anthropologies. That is why Protestants argue for faith alone, and Catholics argue for faith seeking understanding. This Catholic approach to faith seeking understanding is beautifully illustrated in the traditional "proofs" for the existence of God.

Meaning is at the heart of the traditional proofs for the existence of God. The following three statements, understood in a broad sense, are at the heart of all the traditional proofs: 1) Nature is not able to establish its own sense of meaning; 2) Nature must somehow be meaningful; 3) Because nature must somehow be meaningful and because nature is not able to establish its own sense of meaning, then there is a need for a Being which can give meaning to nature. Below are some of the most famous passages that exemplify the traditional proofs of God's existence.

Natural Revelation: By the Light of Human Reason

Scriptural Insights: Wisdom 13:1-5; Romans 1:19-20; Psalm 62:1

Wisdom 13:1-5, NJB
[Ignorant] are all who are unaware of God, and who, from good things seen, have not been able to discover Him-who-is, or, by studying the works, have not recognized the Artificer. Fire, however, or wind, or the swift air, the sphere of the stars, impetuous water, heaven's lamps, are what they held to be the gods who govern the world.

If, charmed by their beauty, they have taken these for gods let them know how much the Master of these excels them, since it was the very source of beauty that created them. And if they have been impressed by their power and energy, let them deduce from these how much mightier is he that has formed them, since through the grandeur and beauty of the creature we may, by analogy, contemplate the Author.

Romans 1:19-20, NJB
For what can be known about God is perfectly plain to them,
since God made it plain to them; ever since the creation of the world,
the invisible existence of God and his everlasting power have been
clearly seen by the mind's understanding of created things.

Psalm 62:1—Need for Rest

"In God alone there is rest for my soul." We were created for God, and thus when we deny God his place in our life, we end up empty.

Many seek to fill this emptiness with power, fame, money, sex, drugs, and so forth. All these things can never fill the emptiness at the core of one's being. They are simply momentary bandages, at times terribly unhealthy bandages.

For example, a man buys his dream car and is filled with enthusiasm. At first he makes sure to park his car in a safe place and makes sure it is kept clean and well running. But as time goes by, what was once appealing loses its appeal. The scratch, the dirt, the very model of the car loses its attraction. Something nicer and newer attracts the eye. One abandons one's dream car, and seeks to buy the new dream car.

It is like a little child who plays with his brand new toy non-stop for the first week. After the first week the toy is nowhere to be found.

Likewise, those that seek fame and acquire it, find momentary satisfaction, but then they find fame empty. They thus seek to acquire more fame so as to recapture some sense of satisfaction.

Others seek wealth, and find momentary satisfaction, but then find themselves needing more. And the pattern goes on. It is an unending, destructive and empty cycle.

Only in God is there rest. Only in God do all of the world's goods find their proper place within a life of fulfillment, a life at peace with oneself and the world. When God fills the emptiness in one's life, then money, fame, chastity, and all the wonderful gifts that the world offers has its authentic preciousness.

It is for this reason that the famous 17th century mathematician, Blaise Pascal, could say that to believe in God is a win win situation and to reject the belief in God is a lose lose situation. One who believes in God lives a happy life, and when one dies, one is rewarded with the gift of eternal bliss in heaven. If the believer ends up wrong, however, it is still a winning situation, for such a person's life on this earthly journey was lived out in happiness and meaningfulness.

On the other hand, the atheist or non-believer is in a lose lose situation. He or she lives a miserable life on this earthly journey chasing after ghosts or gods of fame, power, drugs, and all forms of debauchery. Life becomes a slow debilitating journey into disintegration and death. If the atheist is right and there is no God, then the atheist has lived a miserable life filled with at best momentary satisfactions, and when he or she dies, then that is it—there's nothing more. If, on the other hand, the atheist or non-believer is wrong, then not only does he or she end up living a miserable life here on earth, but when death comes, the eternal pain of hell awaits. Thus, the atheist or non-believer is in a lose lose situation.

Only in God is our soul at rest!

John Henry Cardinal Newman's Argument from Conscience

If, as is the case, we feel responsibility, are ashamed, are frightened, at transgressing the voice of conscience, this implies that there is One to whom we are responsible, before whom we are ashamed, whose claims upon us we fear. If, on doing wrong, we feel the same tearful, brokenhearted sorrow which overwhelms us on hurting a mother; if, on doing right, we enjoy the same sunny serenity of mind, the same soothing, satisfactory delight which follows on our receiving praise from a father, we certainly have within us the image of some person, to whom our love and veneration looks, in whose smile we find happiness, for whom we yearn, toward whom we direct our pleadings, in whose anger we are troubled and waste away. These feelings in us are such as require for their exciting cause an intelligent being: we are not affectionate toward a stone, nor do we feel shame before a horse or a dog; we have no remorse or compunction on breaking mere human law; yet, so it is, conscience excites all these painful emotions, confusion, foreboding, self-communication, and on the other hand it sheds upon us a deep peace, a sense of security, a resignation, and a hope, which there is no sensible, no earthly object to elicit. The wicked flees when no one pursueth; Why does he flee? Whence his terror? Who is it that he sees in solitude, in darkness, in the hidden chambers of the heart? If the cause of these emotions does not belong to this visible world, the Object to which his perception is directed must be Supernatural and Divine; and thus the phenomena of Conscience, as a dictate, seeks to impress the imagination with the picture of a Supreme Governor, a Judge, holy, just, powerful, all-seeing, retributive, and is the creative principle of religion, as the Moral Sense is the principle of ethics (An Essay in Aid of A Grammar of Assent, Westminster Classic, 1973, 109-110).

For John Henry Cardinal Newman there is within every person an inner guide, judge, governor, or principles which helps one discern between that which is good—according to right reason—and that which is evil. This inner guide finds its origin in God and is called "conscience." Our human experience may not recognize the origin of these principles, God, but it is aware of some inner pull tugging at the heart of its most inner being on moral issues (*All cultures throughout the world and history have had a sense of right and wrong which cannot be explained by the simple learning of cultural mores, socialization or inculturation*).

Similar to John Cardinal Henry Newman's argument from conscience is the argument from happiness. It goes as follows:

We were created with an emptiness inside us, an emptiness that only God can fill. Some feel that wealth or money will give them happiness, but the happiness lasts for a short while only, then they chase after more money. Some feel that power or fame will make them happy, but the happiness lasts for only a short while, then they chase after more fame and more power. Some feel that promiscuous sex will be the source of their happiness, but the happiness lasts a short while and they seek more and more. Some feel that drugs will help them cope with life and bring them happiness, but the happiness is short lived, and they seek a bigger and bigger high. Some feel the perfect spouse or the perfect friend will give them happiness, but the happiness lasts for a short while only. The list is unending of what people attempt to fill their souls with.

There was a man by the name of Augustine who tried everything under the sun to capture happiness amidst the unavoidable sufferings of life. He only found it when he found God: "Only in God will my soul rest." He became Saint Augustine.

There are two realities we deal with every day of our lives. On the one hand we desire a life of happiness, peace and contentment, and on the other hand we recognize that life comes with suffering. How do we reconcile these two realities?

As Christians we believe that Christ reconciles these two realities; that in Christ, the Reconciler, we can have a taste of happiness, a taste of peace and contentment, even amidst a life that inevitably and unavoidably comes with suffering, with trials and tribulations.

Holiness is a voyage toward light, happiness and peace. It is a journey toward "wholeness" and toward authentic "humanness."

We all thirst and ache for meaning and purpose: the spiritual life is the only thing that provides the means for attaining this purpose and meaning.

St. Thomas Aquinas: The Five Ways

The first and most obvious way [to prove the existence of God] is based on change. We see things changing. Anything that changes is being changed by something else... This something else, if itself changing, is being changed by yet another thing; and this last change by another. Now we must stop somewhere, otherwise there will be no first cause of the change, and, as a result, no subsequent causes. (Only when acted upon by a first cause do intermediate causes produce a change...). We arrive then at some first cause of change not itself being changed by anything, and this is what everybody understands by God.

The second way is based on the very notion of cause. In the observable world causes derive their causality from other causes; we never observe, nor ever could, something causing itself, for this would mean it preceded itself, and this is not possible. But the deriving of causality must stop somewhere; for in the series of causes an earlier member causes an intermediate and the intermediate a last (whether the intermediate be one or many). Now eliminate a cause and you also eliminate its effects: you cannot have a last cause, nor an intermediate one, unless you have a first. Given no stop in the series of causes, no first cause, there will be no intermediate causes and no last effect; which contradicts observation. So one is forced to suppose some first cause, to which everyone gives the name of God.

The third way is based on what need not be and on what must be.... Some of the things we come across can be but need not be, for we find them springing up and dying away, thus sometimes in being sometimes not. Now everything cannot be like this, for a thing that need not be, once was not; and if everything need not be, once upon a time there was nothing. But if that were true there would be nothing even now, because something that does not exist can only be brought into being by something already existing. If nothing was in being nothing could be brought into being, and nothing would be in being now, which contradicts observation. Not everything therefore is the sort of thing that needs not be; some things must be, and these may or may not owe this necessity to something else. But just as a series of causes must have a stop, so also a series of things which must be.. One is forced to suppose something which must be, and owes this to nothing outside itself; indeed it itself is the cause that other things must be. This is God.

The fourth way is based on the gradation observed in things. Some things are better, truer, more excellent than others. Such comparative terms describe varying degrees of approximation to a superlative... Something therefore is the truest and best and most excellent of things, and hence the most fully in being.... Now when many things possess some property in common, the one most fully possessing it causes it in the others.... Something therefore causes in all other things their being, their goodness, and whatever other perfection they have. And this is what we call God.

The fifth way is based on the guidedness of nature. Goal-directed behavior is observed in all bodies obeying natural laws, even when they lack awareness. Their behavior hardly ever varies and practically always turns out well, showing that they truly tend to goals and do not merely hit them by accident. But nothing lacking awareness can tend to a goal except it be directed by someone with awareness and understanding.... Everything in nature, therefore, is directed to its goal by someone with understanding, and this we call God (Summa, 12-14).

The key points to Aquinas' arguments are as follows: 1) There is change in the world and change requires something or someone to cause this change. 2) There is something that causes other things in the world. 3) There is order in the world (the very laws of physics prove this).

These points lead to essentially the following questions: Do things that change always have a cause for the change? Do causes of things cause other things to change? Do things that are in order have someone or something to put them in order? What does our experience teach us?

When trying to understand the mystery of change, of causes, of order, of degrees of approximation, one is left with either an unending cycle (or what philosophers call an infinite regress) of changes and causes which have no beginning, no first cause, no prime mover, no orderer to order, or one is left with a first cause of change, a prime mover, an orderer to order—God.

Human experience teaches us that it is more likely that "things" have a beginning, for all we know in life is marked by a beginning. We call this beginning without a beginning, the source of all beginning, the first cause, the prime mover, the source of all order, "existence itself," the "I AM," God.

The Big Bang Theory

The "big bang theory" argues that the beginning of the universe had its origins in an extremely compact, dense, and hot mass, a mass of primordial matter made up of protons, neutrons, and electrons in a sea of radiation. The expanding and cooling universe developed into gravitational, electromagnetic, strong and weak nuclear forces, and what would eventually become our modern vision of the universe.

Yet the great flaw of the "big bang theory" is that it cannot explain what happened "before" the big bang. Who caused it?

In reference to the "big bang theory," some have proposed the theory of the oscillating universe. This theory argues that after the "big bang" the universe expanded and will continue to expand to a point where it can no longer expand (due to the universe's density or concentration of mass); once this point of expansion has reached its capacity, the universe's expansion will eventually stop and the universe will begin contracting until it collapses back to its original primordial state, preparing to explode and expand again. Thus the universe, according to this theory, is seen as continually expanding and contracting, continually oscillating.

One is left with the reality that either the universe always existed, oscillating, or God created it. One is left with either an uncreated primordial state of matter (*ylem*) or a God who created this primordial state of matter which eventually led to a cosmic explosion.

Either God is the creator of the universe or the universe always existed. Yet human nature shows us that we have no knowledge of anything that does not have a beginning. Everything we know and everything we experience has a beginning to it. Given this, is it not more likely that God always existed rather than the universe? After all, God is "being in itself," "subsistent existence," the great "I AM."

The Mystery of Miracles

Miracles, which occur in all religions, can always be explained in rational terms. Despite this, they leave every explanation subject to questions. These questions leave a person to reflect on that which can often not clearly be explained.

In the Catholic tradition a saint is ordinarily canonized after two miracles. A person, for example, may be suffering from an incurable and irremovable tumor. The x-rays and all the scientific evidence may point to eminent death. The person calls upon the intercession of a particular holy person who has died, with the hope of a cure. Suddenly, what once was there is no longer there: the tumor which once appeared on the x-ray has disappeared.

For the Catholic this is a miracle, for a skeptical scientist this is an unexplainable phenomenon. An unbelieving scientist would argue that the human body is a complex and not fully understood reality capable of acting in ways that are often unexplainable to the mind. For such a person, something puzzling and contrary to the natural progress of a disease has occurred. Yet this puzzling, unexplainable, phenomenon cannot but make the unbeliever question: How did this occur? Are miracles possible? What is the cause of this mystery?

The Life Principle

I once saw a man die after being hit by lightning. When the coroner examined him, they said that the lightning disrupted his heart rhythm and he died. His body showed no damage from the lightning, except for the fact that it caused the heart to palpitate and stop.

The question arises, what makes a body alive. Given all the same parts, in the same workable condition, what makes a living body different from a dead body? Clearly there is a life principle involved. We call that life principle the soul. But whatever one calls it, the question must be asked: Where does this life principle come from and where does it go after death? If God is not the source and sustainer of this life principle, then what is?

Converging Arguments

The above arguments are not meant to prove the existence of God in the same way that science proves a hypothesis. They are not proofs in the manner of the "natural sciences." Rather, the "proofs" here must be seen as "converging and convincing arguments" that call one to question the plausibility for the existence of God.

Predestination or providence?

If the human race does not have the power of a freely deliberated choice in fleeing evil and in choosing good, then men are not accountable for their actions, whatever they may be. That they do, however, by a free choice, either walk upright or stumble, we shall now prove. God did not make man like the other beings, the trees and the four-legged beasts, for example, which cannot do anything by free choice. Neither would man deserve reward or praise if he did not of himself choose the good; nor, if he acted wickedly, would he deserve punishment, since he would not be evil by choice, and could not be other than that which he was born. The Holy Prophetic Spirit taught us this when he informed us through Moses that God spoke as follows to the first created man: 'Behold, before your face, the good and the evil. Choose the good.'

Justin Martyr (ca. 100)
First Apology, 43

There are some Christian groups that believe in a strict form of predestination. They believe that God predestined some people to heaven and some people to hell. People consequently have no free will to respond to

God's grace. They are either members of the elect or members of the damned.

This extreme view of predestination is rejected by the Catholic Church, for it is a rejection of a human person's ability to love God. How can one truly love if one is predestined to love? That is not love. That is a robot programmed to do what the programmer has asked him or her to do.

True love implies freedom. One loves because one makes a decision to love. A Catholic makes a decision to love God in response to God's showering grace. Strict predestination, therefore, is rejected by the Catholic Church.

The Catholic Church believes in God's providential will. Providence can be seen as a reality somewhere in between extreme predestination and extreme free will. That is, somewhere in between a strict predestination where some are predestined from birth to heaven or hell, and an extreme understanding of free will where God has no say in the world. Providence is the belief that God has a predestined plan for the world and that he knows everything that will take place; he even knows in advance the free will decisions we will make, for these free will decisions are part of God's divine plan.

Free will is affirmed throughout the Scriptures within the scope of God's plan (Gn. 4:7; Dt. 30:19; Sir. 15:14; Prov. 1:24; Is. 5:4; Ez. 18:23; Mt. 23:37; Lk. 13:34; Acts 7:51; Heb. 12:15; 2 Pet. 3:9; Rev. 20:4).

Clement of Rome, the friend of the apostle Peter, in his *Letter to the Corinthians* affirms this Catholic position:

> *Let us look back over the generations, and learn that from generation to generation the Lord has given an opportunity for repentance to all who would return to Him (7).*

Is salvation assured?

"Are you saved?" is the question that is often posed by all kinds of Protestants. As Catholics we can say that at this moment I am saved (cf. Rom. 8:24; Eph. 2:5, 8; 2 Tim. 1:9; Tit. 3:5), but because of the gift of free will, I can in the future deny Christ and lose the salvation that was gifted to me (cf. Phil. 2:12; 1 Pet. 1:9; Mt. 19:22; 24:13; Mk. 8:3-5; Acts 15:11; Rom. 5:9-10; 13:11; 1 Cor. 3:15; 5:5; Heb. 9:28).

For some Christians, all that is required for salvation and its continual assurance is the acknowledgment of Christ as one's personal Lord and Savior (Acts 2:21;19:15, 28; 19:17; Rom. 10:13; 1 Cor. 6:11; 2 Thess. 1:12; 2:19; 1 Jn. 6:11; 5:13). This is what it means, for them, to be "born again" (Jn. 3:3-5). Once this step has been taken the person is saved and can never lose his or her salvation. Guided by God this person will from this moment on live a good life, and if he or she should fail at some moment to live a good life, then the Holy Spirit will come to punish, purify, and to return that person back to wholeness. No matter what, one is saved and can never lose that salvation. It is assured!

Some would point out that if one's life exemplifies great sinfulness, then one really did not accept Jesus Christ as his or her Lord and Savior. On the surface they may have appeared to, but in their soul they did not.

Given the above Scripture quotations, this may appear a convincing explanation for the assurance of one's salvation. Yet the above Scripture quotes do not make any such claim of assured salvation upon the simple proclamation of Jesus as Lord and Savior. These quotes must be taken within the context of the whole of the Bible. When this is done, then the above Scripture quotes can be properly understood.

Basing itself on the correct interpretation of the Scriptures and the constant teaching of the Church, the Church has always affirmed that salvation is conditional. (The belief in the absolute assurance of one's salvation is a novel position that can be traced to the heresies of Gnosticism and 16th century Protestantism). Catholics believe that one's salvation is dependent on one's constant "yes" to God's grace, to God's call. To say that one is assured of salvation by one act, as some Protestants argue, is to essentially say that once one has proclaimed Jesus as Lord and Savior, one's free will has been lost, since one cannot reject God from that point on.

The question must be asked in such a case: "How can one truly love God if one is assured of salvation?" Love implies a free will. Love is a decision, a free decision. To deny any future decision is to eliminate the capacity to love God. We would be saved robots waiting to die and enter into heaven. Where would the virtue of hope be?

For the Catholic, then, one needs to freely choose to respond to God's grace at every moment. One does not lose his or her free will after what some call being "born again." Although these people may argue that one's free will is not lost, the obvious philosophical consequence of saying that one is assured of faith is to claim that one has no more say in the future of one's eternal destiny.

As Catholics, we argue that our salvation is dependent on the state of our mortal soul at the moment of our death (cf. Mt. 25:31-46). A person that dies in the friendship of God, in a state of grace, is granted the rewards of eternal heaven with God and the saints and angels. The person who dies in mortal sin (1 Jn. 5:16-17) will reap what they have sown, eternal damnation.

For Catholics there is a distinction made between redemption and salvation. Jesus has redeemed the world by his blood; he has restored our friendship with God. But redemption is not the same as salvation.

Salvation presupposes redemption, but is distinguishable from redemption. Christ opened the gates of heaven for us, delivered us from sin, and restored humanity to the life of grace by the redeeming act on the Cross.

We in turn must respond to the redemption won for us. We must respond to the engulfing grace he has released upon us (Phil. 1:6; Heb. 13:20-21). The gates are open, but one must choose to enter through those gates (Rom. 2:3-8; 5:9-10; 3:1-13; 3:19-31; 11:22; 13:2; 1 Cor. 1:8; 3:12-15; 4:3-5; 6:9-11; 9:27; 10:11-12; 13:1-3; 15:1-2; 2 Cor. 2:15; 5:10; 13:5; Gal. 5:13-21; 6:8-9; Eph. 2:8-10; Phil. 2:12; 3:7-16; Heb. 10:26-29; 1 Tim. 5:3-8; 2 Pet. 1:1-11; 2:20-21; 1 Jn. 1:5-10; 2:1-11; 3:7; 3:10-17; 3:21-24; 4:20-21; 5:1-5; Mt. 7:21; 19:16-21; 25:31-46; Rev. 2:23; 22:12-15).

If salvation is assured why would we have to be careful and pray for strength against temptations (Mt. 26:41; Mk. 14:38; Lk. 22:46; Gal. 6:1)? Why would one have to train oneself like an athlete for fear of losing one's salvation (1 Cor. 9:27)? If salvation is assured, why do we need to "persevere" (Mt. 24:13; 2 Tim. 2:12)? If salvation is assured, why would we need to do penance (Mt. 3:8; Acts 2:38; 8:22; 2 Cor. 7:10)? If salvation is assured why would we need to be judged by the Lord (1 Cor. 4:4-5; 2 Cor. 5:10)? If salvation is assured why would we be concerned about being paid "according to our works" (Rom. 2:6) or being paid according to our "conduct" (Mt. 16:27)? If salvation is assured, why would we need to "remain in his kindness" for fear of being "cut off" (Rom. 11:22)? If salvation is assured, how can one be in the process of "being saved," or "perishing" (2 Cor. 2:15)? If salvation is assured, why are we called to "test ourselves" and fear the failing "of the test" (2 Cor. 13:5)? If salvation is assured, why must we "work out our salvation with fear and trembling" (Phil. 2:12)? If one is assured of salvation, why bother with religious duties and moral obligations (cf. 1 Tim. 3:8)? If salvation is assured, why bother follow in Jesus' footsteps (1 Pet. 2:21)? If salvation is assured, why acknowledge our sinfulness (1 Jn. 1:5-10). If salvation is assured, why bother to follow the commandments (1 Jn. 2:1-11; Jn. 14:21; Mt. 19:17)? If salvation is assured, why is crying "Lord, Lord" insufficient for entering the kingdom of heaven (Mt. 7:21)? If salvation is assured how do we explain these words from Jesus: "If anyone wishes to come after me, he must deny himself and take up his cross daily and follow me. For whoever wishes to save his life will lose it and whoever loses his life for my sake will save it" (Lk. 9:23-24).

If one was assured of salvation, then faith would not have a future goal? Yet Peter reminds the faithful to persevere during times of trial for they are achieving in this process "faith's goal, salvation" (cf. 1 Pet. 1:6-9). Or as Paul states: "I continue my pursuit toward the goal, the prize of [salvation]" (Phil. 3:14).

One must avoid the sin of presumption, the sin that boasts in a false sense of assured salvation (Jms. 4:13-16). One must remember the words of Paul who reminds us to "work with anxious concern to achieve one's salvation" (Phil. 2:12), and to let no one "think he is standing upright...lest he fall" (1 Cor. 10:12).

The *Didache, The Teaching of the Twelve Apostles*, 16, reminds us that we need to "endure in our faith in order to be saved." And the *Epistle of Barnabas* (ca. 96) reminds us that salvation can be lost at any moment:

> *Let no assumption that we are among the called ever tempt us to relax our efforts or fall asleep in our sins; otherwise the Prince of Evil will obtain control over us, and oust us from the kingdom of the Lord* (*Barnabas*, 4, trans. Walter Mitchell in *Early Christian Prayers*, Chicago: Henry Regnery Co., 1961).

The view that one could be assured of salvation has always been a heretical teaching. The Catholic Church was embroiled for centuries with the heresy of Christian Gnosticism—with its "Great Silence," its aeons, its demiurge, with its elect, with its certainty of salvation (for the *pneumatikoi* or *psychikoi*). Heresies never really go away, they just recycle themselves.

To argue for the assurance of salvation would be to make grace "cheap," as Dietrich Bonhoeffer would say. It would be an insult to the majesty of God.

What about this faith and works?

> *The quality of holiness is shown not by what we say but by what we do in life.*
> Gregory of Nyssa (d. 394)
> PG 46, 262

While it is true that we are justified by faith (Acts 13:39; Rom. 1:17; 3:20-30; 4:5; Gal. 3:11); we are not justified by faith alone (Jms. 2:14f). Let us look at what the Bible says:

> *What good is it, my brothers and sisters if you say you have faith but do not have works? Can faith save you? If a brother or sister is naked and lacks daily food, and one of you says to them, 'Go in peace, keep warm and eat your fill,' and yet you do not supply for their bodily needs, what is the good of that? So faith by itself, if it has no works, is dead.*
>
> *But someone will say, 'You have faith and I have works.' Show me your faith apart from your works, and I by my works will show you my faith. You believe that God is one; you do well. Even the demons believe and shudder. Do you want to be shown, you senseless person, that faith apart from works is barren? Was not our ancestor Abraham justified by works when he offered his son Isaac on the altar? You see that faith was active along with his works, and*

faith was brought to completion by the works. Thus the scripture was fulfilled that says, 'Abraham believed God, and it was reckoned to him as righteousness,' and he was called the friend of God. You see that a person is justified by works and not by faith alone.... For just as the body without the spirit is dead, so faith without works is also dead (Jms. 2:14-24, 26, NRSV).

This passage has been a stumbling block for Protestants from the very beginning. Martin Luther, the first Protestant, wanted to drop the book of James from the New Testament. He called it a "straw letter." It was only after much opposition that he reluctantly left the book in the Bible. This passage is a clear challenge to Protestant theology.

In Matthew 7:21 we read: "Not everyone who says, 'Lord, Lord,' will enter the kingdom of heaven, but the one who does the will of my heavenly Father."

And in Matthew 25:41-46 we read:

[Jesus] will say to [those] on his left, 'Depart from me, you accursed, into the eternal fire prepared for the devil and his angels. For I was hungry and you gave me no food, I was thirsty and you gave me no drink, a stranger and you gave me no welcome, naked and you gave me no clothing, ill and in prison, and you did not care for me.' Then they will answer and say, 'Lord, when did we see you hungry or thirsty or a stranger or naked or ill or in prison, and not minister to your needs?' He will answer them, 'Amen, I say to you, what you did not do for one of these least ones, you did not do for me.' And these will go off to eternal punishment, but the righteous to eternal life.

It is true that we are **not** saved by our works (Eph. 2:8-9). The Church has always believed this. In fact, the heresy of Pelagianism which argued that one could work out one's salvation was condemned in the fifth century.

What Catholics argue is that salvation implies works (Jms. 2:20, 22). That is, one cannot be saved by faith alone. Martin Luther is the one that inserted the word "alone" after faith in his translation of the Scriptures. He did this in Romans 3:28 and Galatians 2:16. Notice that in any scholarly translation "alone" does not appear after the word "faith" in Romans 3:28 and Galatians 2:16.

Salvation for Catholics implies faith and works. Authentic faith always implies the fruits of that faith, works. And authentic holy work always implies a source for that holy work, faith. Faith and works therefore cannot be separated. John Chrysostom's (ca. 344) commentary on the Gospel of John (31, 1) illustrates this point succinctly:

He that believes in the Son has everlasting life. Is it enough then to believe in the Son in order to have everlasting life? By no means! Listen to Christ declare this himself when he says, 'Not everyone who says to Me, Lord, Lord! shall enter the kingdom of heaven'; For if a man believe rightly in the Father and in the Son and in the Holy Spirit, but does not live rightly, his faith will avail him nothing toward salvation.

Maximus the Confessor wrote: "faith must be joined to an active love of God which is expressed in good works." (Cf. *Centuria, cap.* 1, 30-40: PG 90, 967). "If I have all the faith in the world, but am without love, I gain nothing" (cf. 1 Cor. 13:1-3). It is for this reason that the Scriptures remind us that we will be rewarded according to our works: "None of those who cry out, 'Lord Lord', will enter the kingdom of God but only the one who does the will of my Father in Heaven" (Mt. 7:21). "If you wish to enter into life, keep the commandments" (Mt. 19:17-18). "The one who holds out to the end is the one who will see salvation" (Mt. 24:13). "Work with anxious concern to achieve your salvation" (Phil. 2:12). "The just judgment of God will be revealed when he will repay every man for what he has done" (Rom. 2:6). "He will receive his wages in proportion to his toil" (1 Cor. 3:8). "It is not those who hear the law who are just in the sight of God; it is those who keep it who will be declared just" (Rom. 2:13). "[We are saved] by faith, which expresses itself through love" (Gal. 5:6). "[We are] created in Christ Jesus to lead the life of good deeds" (Eph. 2:10). (also cf. Mt. 25:34-36; Lk. 6:27-36, 46-49; Rom. 8:25; 11:22-23; Col. 3:23f; Heb. 10:24-29; Jms. 1:22-25; 2:14-26; 2 Pet. 2:20-21; 1 Jn. 3:7; 5:3; 2 Jn. 8; Rev. 22:12)

Polycarp, a disciple of John, writing on the importance of good works, declares:

When it is in your power to do good, withhold not, because alms deliver from death. All of you be subject to one another, having your behavior blameless among the Gentiles, that by your good works you may receive praise, and the Lord may not be blasphemed in you (10).

The *Epistle of Barnabas* (ca. 96) likewise affirms that

everyone will be recompensed in proportion to what he has done (4). Remember the day of judgment day and night, and seek out every day the faces of God's holy people, either laboring in speech by exhorting others and trying to save souls by what you say, or by working with your hands for the remission of your sins (Epistle of Barnabas, 18:19, trans. Alun Idris Jones in Promise of Good

Things: The Apostolic Fathers, New Rochelle: New City Press, 1993).

Authentic faith implies authentic works, and authentic holy works implies authentic faith.

Blessed Mother Teresa of Calcutta, a saint in our own time, exemplified this inseparable nature between faith and works when she stated:

> *The fruit of silence is prayer.*
> *The fruit of prayer is faith.*
> *The fruit of faith is love.*
> *The fruit of love is service.*

Let us never forget to bear the good fruit that comes from our obedience to faith. Let us keep in mind the words of Paul: "I do not run like a man who loses sight of the finish line. I do not fight as if I were shadowboxing. What I do is discipline my own body and master it, for fear that after having preached to others I myself should be rejected" (1 Cor. 9:27).

Faith operates through love, a work (cf. Gal. 5:6). We will receive our reward from God according to our grace filled works (Rm. 2:6), for faith operates through love, a work (cf. Gal. 5:6).

Let us never commit the sin of presumption. Let us follow Paul's teaching which reminds us to "work out...[our] salvation with fear and trembling" (Phil. 2:12).

Why do Catholics believe that so-called "non-Christians" can be saved?

Joseph of Arimathea...was a disciple of Jesus, though a secret one...(Jn. 19:38).

Joseph of Arimathea was an anonymous follower of Christ. What lesson can we learn from him?

Mohandas Gandhi, the great Indian leader, was killed by gunfire in 1948 by an assassin's bullet. As Gandhi fell to the ground, he placed his hand on his forehead. This was quite significant, for in the Hindu culture, one indicates forgiveness by such a gesture. Gandhi, before his death, had forgiven his assassin. Is this not a holy act? Does this not require grace? Is one not saved who performs such an act?

The Catholic Church affirms that Christ is the way and the truth and the life and that no one goes to the Father except through the Son (Jn. 14:6), and consequently through his Body, his Bride, the Church. All salvation therefore comes from Christ and his Body the Church (1 Cor. 12:12f; 2 Cor. 11:2; Rom. 12:5; Eph. 1:22f; 5:25, 27; Rev. 19:7).

Lumen Gentium 14, in acknowledging Mark 16:16 and John 3:5, affirms the following:

> *Basing itself on Scripture and Tradition, the Council teaches that the Church, a pilgrim now on earth, is necessary for salvation; the one Christ is the mediator and the way of salvation; he is present to us in his body which is the Church. He himself explicitly asserted the necessity of faith and Baptism, and thereby affirmed at the same time the necessity of the Church which men enter through Baptism as through a door. Hence they could not be saved who, knowing that the Catholic Church was founded as necessary by God through Christ, would refuse either to enter it or to remain in it.*

Faith, which implies holy works, baptism, and consequently the Church are necessary for salvation.

The Church, however, makes adamantly clear that there are those who "through no fault of their own" who will be saved.

> *Those who, through no fault of their own, do not know the Gospel of Christ or his Church, but who nevertheless seek God with a sincere heart, and,* **moved by grace,** *try in their actions to do his will as they know it through the dictates of their conscience those too may achieve eternal salvation (LG 16).*

Given the above teachings we must ask ourselves, "How do we reconcile these two positions of the Church?"

The key is found in the second passage's key phrase "moved by grace." One who is authentically holy is one who has the gift of grace at the core of his or her being. And since Christ is another word for grace, Christ consequently is the source of salvation for a person of authentic holiness, whether that person is explicitly aware of it or not. Such a person is saved by Christ who is the way and the truth and the life and that person is brought to the Father through the Son (Jn. 14:6). The soul of such a person is one in which *implicit* faith is being experienced. Such a soul makes one a member of the Mystical Body of Christ, the Church.

This reality finds its most beautiful expression in Matthew 25:

> *Just as you did it to one of the least of my brethren, you did it to me.... Come, you blessed by my Father inherit the kingdom prepared for you from the foundation of the world; for I was hungry and you gave me something to drink, I was a stranger and you welcomed me, I was naked and you gave me clothing, I was sick and you took care of me, I was in prison and you visited me.*

Or in the words of St. Anselm:

When we speak about wisdom, we are speaking of Christ. When we speak about virtue, we are speaking of Christ. When we speak about justice, we are speaking of Christ. When we speak about peace, we are speaking of Christ. When we speak about truth and life..., we are speaking of Christ (Ps. 36, 65-66: CSEL 64,123-124).

But one may ask about the necessity of baptism. The Church from the earliest of times has recognized three forms of baptism: baptism by water; baptism by desire; and baptism by blood. Since grace, Christ, is in the soul of a person who through no fault of his or her own has not grasped the explicit proclamation of the Gospel, that person, because he or she is moved by grace, has accepted a baptism by virtue of desire at an implicit level. In other words, if such a person was fully aware of the Gospel message in its explicit form, then that person would have gladly been baptized by a baptism of water.

So people from other religions (i.e., Jews, Muslims, Buddhists, Hindus, etc.) can be saved if they are holy; that is, moved by grace to a life that can be viewed as a continual "yes" to God. They are saved by Christ who is the way and the truth and the life and by his Church, his Body, his Bride (1 Cor. 12:12f; 2 Cor. 11:2; Rom. 12:5; Eph. 1:22f; 5:25, 27; Rev. 19:7).

If this is so, some may argue, "What's the point in evangelizing?" Evangelizing is not diminished by recognizing holiness in others. In fact, it is made easier. For the mystery of Christ is already within the person at an implicit level. All we need to do, in evangelizing, is nourish the response to grace in that person until that person comes to an explicit recognition of that which is at the very core of that person's being seeking to be expressed fully.

VIII
MISCELLANEOUS ISSUES

Do Catholics practice idol worship?

In Exodus 20:4-5 we read:

You shall not make for yourself an idol, or any likeness of what is in heaven above or on the earth beneath or in the water under the earth. You shall not worship them or serve them; for, I the Lord your God, am a jealous God...(NASB).

Upon reading such a command from God we would wonder why Catholics would have statues or any art work for that matter.

Again when we interpret the Bible we need to interpret it within the context of the whole Bible. It is in interpreting a passage within the context of the whole Bible that we are able to come to the correct understanding of what is meant by a particular Scripture passage. When taken in the context of the whole, this passage refers to worshiping a "graven image" as a god. In other words, worship which is only due to God is being given to a man-made object. Most Christians today understand this, and virtually all scholars, with the exception of—unfortunately—some ill-informed anti-Catholic writers.

I recently went into a fundamentalist book store and, to my surprise, found statues of angels, rings with images of fishes (a symbol for Christ) and rings with WWJD (the abbreviation for "What Would Jesus Do?"). These are all "graven images." I saw paintings and pictures on the wall for sale and all kinds of cards with all kinds of images on them. And as I approached the counter I saw one of the salespersons showing photographs he had taken of his family on a recent trip. Another person was reading a children's book with pictures in it of Jesus and the apostles. How can they be looking at anything that resembles anything in heaven or on earth? Isn't this forbidden by the entire reading of Exodus 20:4-5?

Obviously, this is not what God meant in forbidding the making of images.

Let us look at the Scriptures. In Exodus 25:18-22 we read where God spoke to Moses and instructed him to do the following:

[You] shall make two cherubim of gold; of hammered work shall you make them, on the two ends of the mercy seat. Make one cherub on the one end, and one cherub on the other end. The cherubim shall spread out their wings above, overshadowing the mercy seat with their wings, their faces facing one another. Toward the mercy seat shall the faces of the cherubim be (RSV).

Isn't this a graven image? There it is right in the Bible! God had commanded the making of statues. In Numbers 21:8-9 we read how God commanded Moses to "make a bronze serpent and mount it on a pole." In the fabrication of the tent cloth covering the "Dwelling" the artisans were commanded to embroider cherubim on the cloth (Ex. 26:1). In the building and furnishing of the temple (1 Kings 6:23-28; 7:23-45) images and carved figures abound—images of cherubim, trees, flowers, oxen, lions, pomegranates, and so on.

Archaeological evidence of the first centuries demonstrates that Jewish synagogues were adorned with murals depicting all sorts of things found in nature. The burial grounds of the Christians, the Catacombs, also illustrated various symbols for Christ as well as various biblical images—the most popular being the woman at the well.

This is no wonder since images, icons, and statues were the books for those who could not read. The ability to read was primarily the domain of the well-to-do, the clerics, the aristocrats and the scholars. The common classes saw stories (i.e., two story churches; that is, two levels of stain glass windows depicting biblical stories) as opposed to reading stories. Preachers would often point to stained-glass windows, icons, frescoes and all forms of art to help the faithful understand the message of the Gospel. Churches were "visual libraries" for the faithful in a time when people could not read.

What is forbidden by the commandment expressed in Exodus 20:4-5 is the worship of anything which is not God. Only God is due worship.

In many ancient pagan cultures it was thought that after a statue of a god was made, the god would come to dwell within or around the object created. So pagans would worship the object for they believed their god was dwelling in the object.

As Catholics, and as most Christians today recognize, we do not see statues or any object as worthy of worship. Statues and other forms of art are simply reminders of the true God we worship. Statues and art help us to move our hearts to love the true God that is not found in any statue or work of art. It is just like a husband who has not seen his wife for a long period of time; he looks at a picture of his wife and his emotions are stirred and comforted in his love for her. He does not love the picture; he loves the person represented by the picture. As Cyril of Alexandria (ca. 429) explains:

> *Even if we make images of pious men it is not so that we might adore*
> *them as gods but that when we see them we might be prompted to*
> *imitate them; and if we make images of Christ, it is so that our minds*
> *might wing aloft in yearning for him (On Ps. 113B (115):16).*

For John Damascene (ca. 645), in Jesus Christ, in the Incarnation, the Son of God has ushered in a new "economy" of images. As he stated:

Previously God, who has neither a body nor a face, absolutely could not be represented by an image. But now that he has made himself visible in the flesh and has lived with men, I can make an image of what I have seen of God...and contemplate the glory of the Lord, his face unveiled (De imag. 1,16: PG 96:1245-1248).

Because of the Incarnation, a new era entered into the world. Images representing Christ and therefore the Gospel would take on a new veracity: The words communicated by the Scriptures are illuminated by the image, and the image in turn is illuminated by the words. (Interestingly enough Paul refers to Jesus as the *ikonos*, the icon, of the living God). The second Council of Nicea (787) stated:

We declare that we preserve intact all the written and unwritten traditions of the Church which have been entrusted to us. One of these traditions consists in the production of representational artwork, which accords with the history of the preaching of the Gospel. For it confirms that the incarnation of the Word of God was real and not imaginary, and to our benefit as well, for realities that illustrate each other undoubtedly reflect each other's meaning (Council of Nicea II (787): COD 111).

Artwork which was common among Christians and Jews would now take on a new and more powerful significance. As the *Catechism of the Catholic Church* states: "Christian iconography expresses in images the same Gospel message that Scripture communicates by words. Image and word illuminate each other (1160)." The word on the written page serves as one means of communication and the image on the canvas serves as another.
Catholics do not worship statues or images. We worship God.

Why relics?

A relic is that which is from a saint or associated with a saint and is intended for the spiritual enrichment of the faithful. There are three classes of relics. A *first class relic* is one that is part of a saint's body. A *second class relic* is one that is a part of the clothing of the saint or something that was used or belonged to the saint during his or her lifetime. A *third class relic* is one that a saint has touched, such as a piece of cloth or other object.
Relics are placed in shrines, reliquaries of churches, and placed in altar stones during the consecrations of church altars.
Relics are intended to stir a person's devotion to living a Christ-like life. They are reminders that living the Christian life is not impossible. If others have been able to live it, we likewise can be comforted in the fact that we too can become saints.
Relics have also been associated with various miracles.

In the Bible we read in 2 Kings 13:21: "Once some people were burying a man... They cast the dead man into the grave of Elisha, and everyone went off. But when the man came in contact with the bones of Elisha, he came back to life and rose to his feet." In Acts 5:15-16 we read, "[People] carried the sick out into the streets and laid them on cots and mats so that when Peter came his shadow could fall on one or another of them...and they were all cured." In Acts 19:11-12 we read, "So extraordinary were the mighty deeds God accomplished at the hands of Paul that when face cloths or aprons that touched his skin were applied to the sick, their diseases left them and the evil spirits came out of them."

A relic is that which is from a saint or associated with a saint and is intended for the spiritual enrichment of the faithful.

Are Catholics pagans?

This is a charge only made by the most uneducated of anti-catholic fundamentalists (I would like to emphasize at this point that most fundamentalists are honest and sincere in their convictions, but the reality is that there are some who are outright hostile for reasons known perhaps only to God).

To charge Catholics with paganism is a charge becoming more and more popular today. What these anti-Catholics do is go to a Catholic practice or belief and then try to search out in the world of pagan belief any similarity they can find, and then say, "You see, these Catholics got this practice from this pagan religion!" This is the ultimate in intellectual dishonesty. It is not of God.

Our God is a God of truth. To try to manipulate something that might have a remote similarity to something that is understood completely in a different way is, to put it as gently as possible, devious. Honest intellectual discussion and disagreement is to be encouraged and cherished but slinging nonsense hoping that some poor uneducated person will swallow it up is a sad statement of a person's belief system. We do not gain members by deception, rather by truth. The truth will set us free.

Let us look at some of the common charges leveled at Catholics. Catholics use incense, lamps, holy days and seasons, blessings, vestments, processions, chants, and so forth. Since these are all found in pagan religions, Catholics must be pagans. I would remind the anti-Catholic to read their Hebrew Scriptures, the Old Testament, and they will find all of these practices in the Word of God—and used by the Chosen People of God!

Often anti-Catholics like to take an aspect of a pagan account and grab a phrase and then say, "There we go again, these Catholics got this from paganism!" For example, they like to say that the title "mother of gods," or a "mother of a god," or "queen of heaven" which is found associated with goddesses like Aphrodite, Venus, Rhea, Cybele, and Ishtar are at the source of the Catholic belief in Mary as the Mother of God. Ishtar particularly fascinates them because she is often depicted with a child at her breast.

Now, if you read or hear this out of context, you end up being deceived. Mary is not the Mother of God in the same sense that the pagans refer to their goddesses. The pagan goddesses were viewed as just that, goddesses. Mary is a human creature of God. She is a creature, the greatest of God's creatures, the Mother of Our Lord and Savior, who was God and man. The pagan goddesses were seen as divine and engaged in all kinds of immorality and evil practices. Mary was pure and a virgin.

What about Rome and its Vestal Virgins? Now we know where nuns come from. The pagan world had married ministers too. Does that make all Protestant ministers pagans?

What about the Christmas tree? Jeremiah 10:2-4 condemns the practice of cutting down trees and bringing them into the home to be decorated. The Christmas tree was associated with the pagan myths linked to the winter solstice. For the Christian, however, this pagan belief was detached from its pagan origins and Christianized. The natural triangular shape of the tree became symbolic of the Triune nature of God, God the Father, the Son, and the Holy Spirit. But what about the condemnation of this practice in Jeremiah?

Many things were condemned or practiced in the Hebrew Scriptures (the Old Testament) that are no longer condemned or practiced in Christianity. For example, certain animals were considered unclean (Lev. 11:1-47). Childbirth, contagious diseases, and a woman's menstrual period were associated with being unclean (Lev. 12:1-8; 13:1-14:57; 15:1-33). Those that blasphemed God were to be executed (Lev. 24: 1-23) as well as those who committed adultery. If one injured another, the same injury was to be inflicted upon the one who did the injuring (Lev. 24:20): "A limb for a limb, an eye for an eye, a tooth for a tooth" (Lev. 24:17-22). There certainly would be a large number of limbless, toothless, and blind people around in our society if we still followed these rules!

And what about those who had deformities? Leviticus 21:18f states: "He who has any of the following defects may not come forward [to offer up an oblation to the Lord]: he who is blind, or lame, or who has any disfigurement or malformation, or a crippled foot or hand, or who is humpbacked or weakly or walleyed, or who is afflicted with eczema, ringworm or hernia." If fundamentalists and pseudo-Protestants, especially Jehovah's Witnesses, oppose the Christmas tree on the grounds of Jeremiah, why are they not consistent in following the prohibitions of Leviticus?

(Remember, Jesus came to fulfill, perfect, and surpass the law; what Jesus affirms we continue to believe and practice; what Jesus perfected, transformed, and surpassed—regarding the Old Law—we believe and practice in its new perfected, transformed, and surpassed way).

Furthermore, many Hebrew practices themselves find their origins in paganism. The practice of converting certain practices into one's belief system is not unique to Christianity. Its precedence is found in the Hebrew Scriptures, in the Old Testament. Isaiah 14:13, Isaiah 27:1, Isaiah 38:9-19, Isaiah 51:9-10, Job 41, Psalm 18:13-14, Psalm 29, Psalm 74:13-14, Psalm 89:9-10, Psalm 93:3-4, Psalm 98:7-8, and Hosea 2:16 all have their origins in Ugaritic or Canaanite myths.

The Israelites also shared with the pagans, particularly the Canaanites, the practice of sacrifice (Lev. 1-8), the use of similar altars (Ex. 27:2; Lev. 9:9; Ps. 118:27; 1 Kgs. 8:31). Both religions had altars of incense. Both had temples made up of a porch, a sanctuary, and an inner sanctuary. Both used memorial markers to mark important places (Gen. 28:18; 35:14; Ex. 24:4). Both placed heavy emphasis on the care and respect of their parents. There are even cases of human sacrifices described in the Scriptures (1 Kgs. 16:34; 2 Kgs. 3:27; Jgs. 11:30-39; 2 Sam. 21:1-9).

Judaism was in no way pagan, any more than Christianity is. Both adapted and Judaized or Christianized certain pagan myths or practices.

Let me turn the table a little. Why do anti-Catholic fundamentalists not claim that the well-accepted belief in the Trinity is not of pagan origin? After all, Mesopotamia had a Triad of Anu, Bel, and Ea, and in terms of Egypt, you can find a mythological system filled with Triads of gods. Yet clearly these are nothing like the Christian understanding of the Trinity. But can you see how someone could misuse this against fundamentalists or Catholics?

What about the idea of a single, supreme god? Aton in Egypt was worshiped as the only supreme god of the heavens. Yet Aton can never be confused with the Judea-Christian God.

What about the idea of a devil? One could look to the god Hades (or Pluto) and find him described as the god of the underworld where no prayers are heard. Is this where we get the devil from? Obviously not! Hades is very different from our understanding of the devil when we examine the myth closely.

What about the idea of a son of god? Pagan mythology is filled with sons of god (i.e., Eros, Cupid—an infant son of god—Mars, Hermes, Mercury, Poseidon, Triton, Neptune, Hephaestus, Vulcan, etc.).

Zeus or the Roman version, Jupiter, is the creator in pagan mythology of humankind and is known as the god of justice and mercy, the protector of the weak, the one who punishes the wicked. He is a King. Apollo is known as the god of truth and the one who overcame the evil serpent. Persephone ate a pomegranate seed and was banished (does that remind you of anything?). Saturn was known as the god of peace. Dionysus died and resurrected and performed miracles. This god was known as the vine. Didn't Jesus refer to himself as the vine and we as the branches? Marduk was called "Lord," the "supreme god," the "creator of the universe," the "creator of all life." Now you better not pray before or after meals since Hestia or Vesta demanded this! You better take that image of the fish off the back of your car, because that's the image of Dagon! What about the story of creation and the flood in Genesis which is predated by an account of creation and the flood in certain Mesopotamian myths (the *Enuma Elish* story and the *Epic of Gilgamesh*)? What about the Emperor Augustus Caesar? He was referred to as a god and the "savior" of the world and the "prince of peace."

If this were not enough, let us look at Christianity's competitor in early Church history, Mithraism:

Mithraism was based on the ancient Persian god of light and wisdom. In the sacred book of writings, the Avesta, Mithra is depicted as the good spirit and ruler of the world. He killed a divine bull, thus creating all plants and animals. He also became known as the god of the Sun, which developed into Sun Worship. Mithraism had many similarities to Christianity, such as the belief in brotherly love, rites of baptism and communion, the use of holy water, the adoption of Sundays and December 25 (the birth of Mithra) as holy days, the immortality of the soul, the last judgment, and the resurrection from the dead.

It is obvious that one can manipulate myths to one's advantage. When we examine the above myths, and I encourage you to do so, you will find that they are quite different. One could never confuse them for our Christian faith. But when we take phrases or aspects of a story out of context, it can surely seem troubling to someone who has never read about these stories, these myths.

Just as you can find the idea in paganism of venerated books that were to be the sole source of truth (*sola scriptura*) and the belief that salvation can be assured of here on this earthly journey in a single act of faith, so too can we find similarities in the other mainline religions of the world. Yet the content of the faith of the various religions is different. Let us always be honest in our debates.

Finally, it is true that Catholics as well as Protestants have adopted practices from other beliefs, but we have purified them and made them in conformity to God's will. For example, rings are exchanged in marriage. That is a pagan practice that was given a Christian meaning. Catholics light votive candles, but unlike their pagan counterparts, the candles that are lit are a symbolic reminder, as the smoke rises up, of our prayers going up to the Father, through the Son, and in the Spirit. Many practices have been Christianized. What was good was kept and Christianized, what was evil was forbidden.

I would like to finish with perhaps the most famous case of Christianizing a pagan practice: the celebration of Easter. Easter was originally a pagan feast of renewal named after the goddess Eastre, the goddess of spring. The Church Christianized this feast by essentially saying to the pagans: "Why do you worship a false god? Why not worship the true God, Jesus Christ. He is the true means of renewal, the true resurrection, the true and eternal spring of life."

What was once pagan now became Christian. How many people worship Eastre, the goddess of spring, today? When you hear the word Easter, you think of one and only one thing, Christ, the true Son of God, the true Savior of the World, the one who died for our sins and was resurrected and ascended into heaven.

All religions have things in common. God has flooded the world with his grace. The very existence of religions or beliefs is the result of one's search for that which is beyond oneself, God. Paganism had a belief in the supernatural, the desire for immortality, for spiritual union, for spiritual guidance. Although the approach was different and imperfect, God's grace cannot be said to have been non-existent in those religions that authentically sought God.

Why do some Catholics believe in the theory of evolution? Are they not contradicting the Scriptures?

There are three competing theories to describe the creation or evolution of man and woman. Some argue for the theory of *Creationism* which maintains that God created man and woman, Adam and Eve, without the necessity of an evolutionary process. There is the theory of *atheistic evolution* which maintains that human life evolved from lower forms to higher forms by a random process. Finally, there is the theory of *theistic [God-guided] evolution*, the belief that God created the world out of nothing and that he guided an evolutionary process from a lower form of life to a higher form of life, until he finally placed an immortal soul into the first human beings.

One, as a Catholic, can believe in a form of *Creationism* or one can believe in *theistic [or God-guided] evolution* as understood here. One cannot however believe in atheistic evolution, for it denies God's creative power and his providential will.

One may wonder how the belief in *theistic [God-guided] evolution* can be believed in terms of the account of creation in the book of Genesis. First and foremost, Genesis is not a historical account of the way the world began, nor the way human life began. All one needs to do is to compare Genesis 1 with Genesis 2:4f. Here, within the first two chapters of Genesis, you find two different accounts of Creation. It is obvious that Genesis was not meant to be a literal historical account of how human beings came into being.

The Book of Genesis is a theological account teaching us that God is the ultimate source of being. He created the world and people out of nothing. He created them good. It is an account of freedom and the cost of using freedom in a negative manner. It is an account of two people, Adam and Eve, who chose to rebel against God and sought to live without God. By their sin they forever distorted the nature of the world. Christ would have to come to save the world from the damage that was caused by the Fall, the "original sin" of Adam and Eve.

Genesis is the Word of God told to a people thousands of years ago about the eternal truths of God, a God of mercy and love, a God of second chances.

Consequently, one, as a Catholic, can believe in a form of *Creationism* or one can believe in *theistic [God-guided] evolution* as understood by the Church.

Are Catholic doctrines invented?

Many evangelicals like to point to a council and say: "You see, this belief only began in such a time." For example, they would argue that the title "Mother of God" was invented at the Council of Ephesus (ca. 431) or Vatican I (ca. 1869-1870) invented Papal infallibility. This of course is an absurdity to any historian or any well-informed Catholic or Protestant. Just because a doctrine is defined specifically does not mean it was not always held to be true. The inscription "theotokos," "God-bearer" is found on the walls of the Catacombs hundreds of years before the Council of Ephesus ever took place.

In terms of Papal Infallibility, Augustine of Hippo (ca. 400) would say in issues of faith and morals the famous phrase: "Rome has spoken; it is settled."

The Popes always had primacy of power: Pope Clement of Rome's *Letter to the Corinthians* (ca. 88) was so respected by the community of Corinth that the letter was almost put into the canon of the New Testament. Anacletus (76-88) was consulted regarding the proper consecration of bishops. Alexander I (105-115) issued the decree that unleavened bread was to be used for consecration; Sixtus I (115-125) decreed the praying of the *Sanctus* and Telesphorus (125-136) the praying of the *Gloria*. Pius I (140-155) issued the decree regarding the proper date for the celebration of Easter. Hyginus (136-140) was asked to squash the heresy of Gnosticism, Anicetus (155-166) the heresy of Manichaeism, Soter (166-175) the heresy of Montanism, and Victor I (189-199) the heresy of Adoptionism, and on and on.

When you study all 265 popes, all 264 successors of Peter, you find without a question that the faith held by the popes became the faith of the Church!

Councils help to clarify a teaching when confusion seems to be harming the belief of the faithful. The Church usually defines a doctrine in a council when there is either hostility to a teaching or confusion over a teaching.

The Church can only teach infallibly what has always been present in the deposit of the faith. It can grow in its understanding of that deposit of the faith and therefore bring about clarification, but that deposit of faith must always remain the same.

A final point: If we were to take the arguments of anti-Catholics seriously, then we would have to argue that Christ's divinity was invented at the First Council of Nicea (ca. 325), that the divinity of the Holy Spirit and the reality of the Trinity was only invented at the Council of Constantinople (ca. 381), and that the Bible was invented at the Councils of Hippo (ca. 393), Carthage III (397) and Carthage IV (419).   How absurd!   But if such anti-Catholics are to be consistent in their argumentation, then they must believe the absurd.

What is a heretic?

A heretic is one that chooses to deviate from the true and authentic teachings of Jesus Christ and his Body, the Church.   One may ask the question, "How is it that some of the earliest saints believed in something different than Catholics believe today?"

When we look at some of the early Fathers of the Catholic Church we see a slow process of understanding the nature of the Trinity (i.e., the relation of the Three Persons, the relationship between Jesus' human nature and divine nature, etc.). These Fathers were pioneers. They were not denying something that was already defined as infallible. They were doing theology. Because of their pioneering work, we have come to a deeper understanding of the faith.

These Fathers of the Church were not heretics.   They were pioneers! Once the Church defined infallibly a dogma of the Church, these Fathers obeyed and accepted these teachings.

A heretic is one that after an infallible teaching has been proclaimed, denies that infallible teaching.   That person is not a pioneer, but a heretic— that is, one that deviates from the truth.   Martin Luther, Ulrich Zwingli and John Calvin are considered heretics, for they denied what became "infallible teaching" at the Council of Trent (ca. 1546).

Today the term is rarely used since it has a malevolent connotation. The reality remains, but the term has been abandoned for the most part.

The Real Jesus!

What would Jesus do? All Christians are called to imitate Christ. Yet in order to understand what Jesus would do, we need to know what Jesus in fact did.   This requires us to study and meditate on the Scriptures, for as St. Jerome so eloquently said: "Ignorance of the Scriptures is ignorance of Christ."

The sad reality is that our modern society has made and portrayed Jesus into a politically correct, limp, unconditionally accepting, superficially affirming, non-judgmental accepter of anything!

Yet this is not Jesus! Jesus unconditionally loves, but he does not unconditionally accept. Unconditional love is based on truth.   To accept someone unconditionally while one is living a lie is acceptance, but not love.

In fact, it is cruelty.

Now some like to say that Jesus taught us not to judge. That is not true. Jesus taught us not to judge a person's soul, but he did not teach us to ignore actions. If he did, then we would never know what was right from wrong.

But I am digressing. Who is the real Jesus we must imitate? Let us look at the Scriptures.

Who said:

Woe to you, Chorazin! Woe to you, Bethsaida! ...It will be more tolerable for Tyre and Sidon on the day of judgment than for you. And as for you, Capernaum...you will go down to the netherworld.... I tell you, it will be more tolerable for the land of Sodom on the day of judgment than for you (Mt. 11:21-24).

Who said this? Jesus did!

Who said:

Woe to you, scribes and Pharisees, you hypocrites. You lock the kingdom of heaven before human beings. You do not enter yourselves, nor do you allow entrance to those trying to enter. Woe to you, scribes and Pharisees, you hypocrites (Lk. 10:13).

Who said this? Jesus

Who said:

But woe to you who are rich, for you have received your consolation. But woe to you who are filled now, for you will be hungry. Woe to you who laugh now, for you will grieve and weep (Lk. 6:24-26).

Who said this? Jesus did!

Who said: "It is easier for a camel to pass through the eye of a needle than for one who is rich to enter the kingdom of God" (Mt. 19:24). Jesus did!

When the woman who was caught in adultery was forgiven by Jesus, he did not say, "Go on and continue in your ways." Rather, he said, "Go and sin no more!"

Who said:

I have come to set the earth on fire, and how I wish it were already blazing! There is a baptism with which I must be baptized, and how great is my anguish until it is accomplished! Do you think that I have come to establish peace on the earth? No, I tell you, but rather division. From now on a household of five will be divided, three against two and two against three; a father will be divided against his son and a son against his father, a mother against her daughter and a daughter against her mother, a mother-in-law against her daughter-in-law and a daughter-in-law against her mother-in-law (Lk. 12:49-53).

Who said this? Jesus.
Who said:

Then he will say to those on his left, 'Depart from me, you accursed, into the eternal fire prepared for the devil and his angels. For I was hungry and you gave me no food, I was thirsty and you gave me no drink, a stranger and you gave me no welcome, naked and you gave me no clothing, ill and in prison, and you did not care for me.' Then they will answer and say, 'Lord, when did we see you hungry or thirsty or a stranger or naked or ill or in prison, and not minister to your needs?' He will answer them, 'Amen, I say to you, what you did not do for one of these least ones, you did not do for me.' And these will go off to eternal punishment (Mt. 25:41-46).

Who said this? Jesus!
Who did this:

Since the Passover of the Jews was near, Jesus went up to Jerusalem. He found in the temple area those who sold oxen, sheep, and doves, as well as the money-changers seated there. He made a whip out of cords and drove them all out of the temple area, with the sheep and oxen, and spilled the coins of the money-changers and overturned their tables, and to those who sold doves he said, 'Take these out of here, and stop making my Father's house a marketplace' (Jn. 2:13-16).

Who did this? Jesus!
Who said:

I know your works; I know that you are neither cold nor hot. I wish you were either cold or hot. So, because you are lukewarm, neither hot nor cold, I will vomit you out of my mouth (Rev. 3:15-16).

Who said this? Jesus!
Jesus is the way, the truth, and the life of all Christians. He came to fulfill, perfect, and infuse the law with his presence, not to abandon it!

This means that Jesus demands of us a response to this inner law of grace, of conscience. This means that Jesus demands a response to the gift of grace, a response of faith, hope, and love, a response of love of God and neighbor.

In order to understand what Jesus would do we need to know what he in fact did and said; otherwise, we will have a distorted image of Christ. Sadly to say, the image of Jesus has been distorted by too many. The image of Jesus has been hijacked by a secular culture that seeks to focus only on a Jesus who fits its agenda of moral relativism.

There is no doubt that Jesus is the faith-filled, hope-filled, charity-filled Lord of lords, King of kings. There is no doubt that Jesus is prudent, temperate, just, and courageous, that he is the humble, generous, patient, peaceful, modest, mild, friendly, and the good and holy Servant of servants. But he is so in the Christian sense, not in the secular sense. As Pope Benedict XVI states: "Love without truth is blind and truth without love is empty."

Let us as Christians not allow the forces of evil to make our Jesus into a politically correct, unconditionally accepting, superficially affirming, non-judgmental accepter of anything!

> *Ignorance of the Scriptures is ignorance of Christ.*
> *St. Jerome*

The health and wealth gospel!

> *It is a poverty to decide that a child must die so that you may live as you wish.*
> *Blessed Mother Teresa of Calcutta*

What Mother Teresa has said about abortion can very well be said about the hungry, the poor and the homeless around the world.

It is so easy for us to harden our hearts to the plight of the less fortunate (cf. Ps. 95). Yet the Gospel and the *Catechism of the Catholic Church* continue to remind us of the need to love our neighbor, and to have a preferential option or love for the poor (cf. Mt. 25:31-46; 5:42; 6:2-4; 8:20; 10:8; Lk. 6:20-22; Mk. 12:41-44; Jas. 2:13-16; 5:1-6; Eph. 4:28; cf. 1 Jn. 3:17) (CCC 2448; *Libertatis conscientia*, 68).

Helping the needy is not as much an act of charity as it is a demand for justice (CCC 2446). It is an act of justice that has always been part of the Church's teachings (cf. Mt. 25:31-46; 5:42; 6:2-4; 8:20; 10:8; Lk. 6:20-22; Mk. 12:41-44; Jas. 2:13-16; 5:1-6; Eph. 4:28; cf. 1 Jn. 3:17). As John Chrysostom (d. 407) explains: "Not to enable the needy to share in our goods is to steal from them and deprive them of life" *(Hom. In Lazaro, 2,5: PG 48, 992)*. Blessings are to be shared.

Isolation is not part of Catholic tradition or spirituality. Christ calls us to be his ears, his eyes, his hands, his feet, and his voice in a world crying for him.

The Church, the body of Christ, demands of us a desire to build up a world where a solidarity of nations can be established to eliminate hunger, poverty, and homelessness (CCC 2438). It demands of us that we aid in the moral, cultural, and economic development of countries (CCC 2438; 2440). This is a grave and unavoidable responsibility for the wealthiest nations (CCC 2439). This is a grave and unavoidable responsibility for each and every one of us who call ourselves Christian: "How can God's love survive in a man who has enough of this world's goods yet closes his heart to his brother when he sees him in need" (1 Jn. 3:17)? May the Lord have mercy on our souls if we remain silent and inactive.

One of the greatest heresies to enter into the life of the Church is the health and wealth gospel. This approach argues that the faithful will remain healthy and grow in the world's riches and goods. It is a sadly unchristian and secular vision of the gospel that has infected too many ecclesiastical communities today.

The Gospel reminds us that we are not worthy of Christ unless we take up our cross and follow him (Mt. 10:38). We can only be heirs with Christ if we "suffer with him" (Rom. 8:16-17). We as Christians are called not only to suffer with him but to suffer for him and to make "up what is lacking" in his sufferings (cf. Phil. 1:28-29; Col. 1:24; 1 Pet. 1:6). In 1 Peter 2:19-21 we are taught: "If you suffer for doing what is good, this is a grace before God. For to this you have been called, because Christ also suffered for you, leaving you an example that you should follow in his footsteps."

And what about wealth? Jesus reminds us, "Blessed are the poor" (Lk. 6:20). He reminds us, "If you wish to be perfect, go, sell what you have and give to the poor" (Mt. 19:21; Mk. 10:21; Lk. 18:22). Jesus warns us, "It is easier for a camel to pass through the eye of a needle than for one who is rich to enter the kingdom of God" (Mt. 19:24; Mk. 10:25; Lk. 18:25).

Jesus is the way and the truth and the life (Jn. 14:6). We are called to imitate this way and truth so that we may have life to the fullest, and life to the fullest does not necessarily mean a life of comfort and wealth!

The New Age

The New Age is simply the recycling of failed religions. It is the simple recycling of animism, paganism, gnosticism, monism, pantheism, gaiaism, occultism, witchcraft and relativism. It is a vision of life whose faith system is preoccupied with self-actualization, channeling, divination, the evolution, reincarnation and transmigration of souls, the astral projection of souls through Yoga, hypnotic rituals, and other methods into "other realms" of reality. The new age movement is all about the shifting of world and personal paradigms.

The new age is an amalgamation of belief systems that people pick and choose from in order to give themselves a sense of security amidst a life of insecurity.   Superstition, animism, paganism, Gnosticism, monism, pantheism, occultism, witchcraft, divination, channeling, magic, materialism, self-centeredness, the worship of self—i.e., the self-actualizing of oneself— the desire for divinity, the desire for reincarnation, the worship of the world or other astral realities is a violation of the First Commandment. The use of crystals for healing, etc., and the denial of the power of mortification and suffering is a violation of the Third Commandment.   The cult of the body, with its recourse to obsessive plastic surgery, invitro-fertilizaton, designer babies, surrogate mothers, cloning, abortion, contraceptives, and even euthanasia is a violation of the Fifth Commandment. The lack of scruples with regard to fornication, adultery, polygamy, open marriages, divorce, homosexual and bisexual acts, masturbation, and pornography is a violation of the Sixth Commandment. The obliviousness to indecency and immodesty is a violation of the Ninth Commandment and the obsession with materialism and the world is a violation of the Tenth Commandment (Cf. Ten Commandments, Ex. 20:1-17).

The New Age Movement is simply an amalgamation of failed philosophies from the past redressed as something new. As it distorted minds in the past, it will again. Heresies and evils never really die; they simply recycle themselves under newer names.

Jesus reminds us: "You shall love the Lord your God with all your heart, and with all your soul, and with all your strength, and with all your mind; and your neighbor as yourself" (Lk. 10:27, NRSV). Anything contrary to this Christian truth is a path to self-destruction, to one's disintegration as a human being.

IX
UNDERSTANDING THE RELIGION OF SECULARISM

Birth pangs of secularism
Nominalism

With the growth of nominalism in the Middle Ages came the birth pangs of secularism. Nominalism argued against the belief in universal essences.

Nominalism argued that intellectual concepts had to correspond to that which was apparent, observable, testable, and in nature. Abstracts, universals, and concepts beyond the sphere of the intellect were therefore non-existent.

The consequences of nominalism would inevitably lead to a preference for subjectivity in thought and attitude and the rejection of abstracts, universals, and absolutes, for if one is trapped within the limits of the mind, then one cannot grasp a God that transcends the limits of concepts and the limits of one's intellectual capacity.

The apple of secularism could not help but be appealing under such a vision of reality.

Descartes and Kant

Rene Descartes and Immanuel Kant were believers in God, but the philosophical systems they would develop would have dire consequences for this belief. Their philosophical systems would continue nominalism's movement of preference for subjectivity over objectivity, immanence over transcendence, philosophy over theology.

Rene Descartes gave subjectivity more fuel by allowing doubt and skepticism to run uncontrolled. He also elevated subjectivity to new heights by his overemphasis on the "real" as that which could only be clearly and distinctly perceived in one's mind.

Immanuel Kant's synthesis of nominalism, Cartesian subjectivity, Leibniz-Wolffian rationalism and Humean skepticism would give rise to what he would refer to as a Copernican revolution in philosophy.

For Kant reality is known only to the extent that it conforms to the structure of the knowing mind. Only that which is experience-able can be known. Things-in-themselves—that is, posited objects or events which are independent of perception by the senses—are not knowable and must be accepted by faith alone.

With Kant we have reached the apex of the revolution. It is but a small leap from faith alone, to no faith at all. It is but a small leap from that which must be conformed to the structure of the knowing mind to relativism and all the modern ethical philosophies of secularism.

Implications of the turn toward the self
Hobbes and egoistic morality

Thomas Hobbes viewed human action as directed toward self-preservation, the seeking of pleasure and the avoiding of pain. All of a person's natural instincts and passions are self-regarding. All that a person does is directed toward doing what is good for the self. A person does not seek to be charitable for its own sake; rather, the person is charitable because he or she receives personal satisfaction from the act of charity.

Whatever is good is that which a person's appetites desire. Whatever is bad is that which a person dislikes or sees as an aversion. These appetites and dislikes or aversions, however, are always seen within the context of self-preservation.

The role of the state and its authority is essential. If the state is not there to guarantee a rule of conduct, then no one is obliged to follow any rule. If rules are enforced, then it is to a person's advantage, to a person's need for self-preservation, to keep the rules.

This inexorably makes morality and secular legality one and the same. What is legal is moral. If abortion is legal, it must be moral!

Hobbes' vision of morality was hedonistic, subjective, and arbitrary in tendency.

Hume

David Hume, while not always consistent in his thought processes, tended to hold the following view regarding ethics: something is good if most people would approve of it as being good. The consensus or majority opinion is what distinguishes the good from the bad. Morally good acts are those which bring the greatest pleasure to the greatest number of people, to the majority. Hume could rightly be considered a forerunner to general utilitarianism.

As opposed to reason directing a person's moral decisions, feelings have priority. In other words, reason is the salve of "approval feelings." Whatever is acceptable to a culture becomes acceptable morally.

Right and wrong become whatever is popular at the time. Slavery was once acceptable in our culture, and therefore was morally permissible. Now slavery is no longer acceptable by the culture, and therefore it is no longer morally permissible. Something is good if most people approve of it as being good.

Hedonistic utilitarianism

Utilitarianism, whose most famous proponents were Jeremy Bentham, John Stuart Mill, and Henry Sedgwick, is a system of ethics based on egoistic and/or general hedonism. The goal of the person is to seek the greatest pleasure or the production of the greatest pleasure and nothing else, and in the case of general hedonism, what counts is the greatest pleasure of the greatest number. Intuition, rational insight, feelings, public opinion, and social and governmental laws have an impact on this quest for pleasure. There is no such thing as that which is intrinsically right or wrong. Right and wrong are simply a matter of the pleasurable consequences that acts bring about. The "pleasure principle" is all that ultimately counts. Nothing but pleasure is good.

Ideal utilitarianism

G.E. Moore is the main proponent of this system. Moore believed that pleasure alone was insufficient as a criterion for what was good. While Moore maintained that there was no absolute proof of an intrinsic good, direct inspection, rational insight or intuition (as opposed to feelings or volitional attitudes) could help one determine a perceived good.

For Moore there are things other than the "pleasure principle" that determine something as good. For example, Moore argues that rational insight and intuition allow for knowledge, virtue, beauty, and personal affection to be perceived as goods. That which is right is that which will bring about the greatest amount of good, and this right is equated with that which is directed by "duty" or what "ought" to be done. Acts are right or wrong according to their consequences and to the existence of the good that they bring into being— the good being determined by insight and intuition.

Spinoza

Benedictus de Spinoza saw philosophy as rational religion enriched by mysticism. Human enlightenment, human freedom, and the human ideal are based on the directing of one's life by natural reason.

For Spinoza, the call of the person is to free himself or herself from bondage. This is done by seeking to understand one's human nature and then seeking to control it. When one is able to explore one's uncontrolled and uncoordinated "active emotions," senses, beliefs, instincts, and "passive emotions" then one can begin to detach oneself from this bondage and grow in self-knowledge and self-control. Rational insight, intellectual development, and intuitive knowledge bring the person to his (her) ideal self and therefore allow him (her) to be in moral conformity with himself (herself), others, and the universe.

The consequence of such a system is that it leads to a sense of right and wrong, of good and bad, based on one's own powers of the intellect, which are subject to change with time. At one time in history the powers of the intellect favored eugenics, today it favors a more moderate form of eugenics called euthanasia. At one time slavery was favored as rational, today it is in disfavor and irrational. The consequence of such a system is the continuation of subjectivity in the modern mindset.

Kant

Immanuel Kant maintained that the moral goodness or badness of an act is based entirely on its "motive" and that the only unconditionally good motive is that of "duty." Since reason is corrupted by the passions, one is bound to live according to "categorical imperatives" such as "thou shall not steal" as opposed to hypothetical imperatives associated with the passions, such as "if I want to be popular, I should not steal."

But what makes something a "categorical imperative"? Only those things which can be universalized as a universal law for all people to follow.

Kant (unlike Spinoza who overemphasizes the power of reason) underestimates the power of reason. Kant was infected with the Protestant notion of "original sin"—the notion that original sin destroyed human nature as opposed to the Catholic position that maintained that original sin wounded human nature. As a consequence, Kant viewed the reasoning faculty as suspect and preferred to stand on the notion of "duty" to the categorical imperative.

The concept that states that we are to act only on those things which can be universalized as a universal law for all people dangerously implants a practically unmitigated subjective dimension into morality, for it detaches the subjective dimension of the person from the objective truth upon which the person should act within his or her subjective circumstances. Just because something can be universalized does not mean it is true or good.

By overemphasizing "motive" Kant failed to recognize that the sense of duty can be different in different people and can lead people into divergent directions. The soldier and the conscientious objector are both prepared to will that the maxims of their actions should become universal law.

Hegel

G.W.F. Hegel argued that the content of moral judgments came from the laws, institutions and customs of a community. The only adequate moral standard to be guided by was that which was in harmony with the social relations in which a person found himself (herself) placed in. Even a person's own conscience cannot go against the demands made upon him or her by the law and the traditions of his or her society. The state is the highest expression of rational morality. Ideal morality is the union of the conscience with the laws and traditions of the state. Ideal morality occurs when a person, from a sense of duty, upholds and conforms his or her life to the rational political institutions and laws of the state.

The problem with Hegel's view is that it fails to recognize adequately the subjective dimension inherent in laws and traditions. It also fails to recognize that law, tradition, consensus, and majority rule in no way assure the good or the right. Hegel's view implies an inability to transcend one's own historical environment.

Evolutionalism/ Progressivism

Evolutionists sought to apply Darwin's theory of evolution to the understanding of morality. They maintained that the individual and/or society was to be directed toward that which assured survival or well-being.

Friedrich Nietzsche maintained this to the extreme. He maintained that which assured survival was moral, no matter how brutish, how cunning. Will and not reason, power and the will to power are the keys to life. One is called to shed one's inhibitions and taboos and to live as one wills.

Evolutionists never found any adequate criterion for social survival or for the nature of "moral obligation." They also failed to adequately explain the timeless and unitary nature of the human person.

This vision of reality is marked by the viewing of every epoch as a succession of improvements. Each epoch is viewed as superior in value and nearer to the truth. Each epoch is viewed as "coming of age," as a new "enlightenment," as a new age where darkness and immaturity are left behind.

The problem with such a world view is that it fails to consider the reality of concupiscence sufficiently and exhibits a false, unwarranted, indiscriminate, enthusiastic optimism for the new. It assumes that free will improves with every epoch, as if free will and collaboration is set in stone, irreversible, in terms of its upward and onward movement. It fails to take into account false prophets who bring regressive and repressive movements into societies, who bring whole cultures into moral blindness, into distorted, wrong ideas and visions.

Ethical Relativity

Right and good are relative terms and describe or express the desires or feelings or preferences of the individual toward actions contemplated by the individual. Moral experience is expressed in a variety of ways. There are no self-evident truths or propositions.

W.G. Sumner argued that right or wrong, good or bad, were simply implanted into us by social pressures. These terms are arbitrary and are only important to the extent they serve the interests of society itself. Whatever is approved or benefits my society is right.

Ethical relativity is based on the view that knowledge is relative to the limited nature of the mind and the conditions of knowing; that is, ethical truths are totally dependent on the individual and/or group. Truth and goods are not absolutes that can be found, but are ever changing concepts.

Relativism shifts a person's vision of the world from a desire for objectivity to an acceptance and comfort with subjectivity. Good and evil are arbitrary decisions made by individuals or by collective groups.

The concepts of transcendence, providence, and the natural law are replaced by an immanent vision of the world where the person seeks to control life, situations, and what is good or bad according to the self. It is a world based on chance occurrences and not on providence.

Relativism has colored the secular and religious climate of the age we live in. Truth and the good have been replaced or dethroned by "aliveness," "dynamism," "operativeness." Truth and the good have been replaced by the historical "fashionableness" of an idea or ideas, with the sociological efficacy of an idea.

The exclusively immanent experience of life comes to the fore in such a view. Truth and good are seen as relative and determined by socially accepted preferences—often under peer pressure—of what is true, good, beautiful, purposeful, meaningful, and so forth.

Only a culture that has been dummied down by comfort, sloth, hedonism, and self-centeredness can fall for such a superficial, schizophrenic, distorted vision of reality.

This has given rise in our time to the concupiscent society where perversion is viewed as normal and religion is viewed as a weakness, a substitute for addiction, an opiate, a projection of self or society.

Relativism by rejecting history makes history meaningless. It makes history devoid of its very nature. History by nature presupposes tradition. History contains the transmission of truths, ideals, cultural treasures, insights, timelessness, and intrinsic values which are realities for every epoch and which require the perseverant clinging to them.

We need to delve deeply into history for God's providence! Relativism diminishes this gift of history.

Pragmaticism/Immanentism

John Dewey maintained that thought is an instrument of practice whereby one molds one's environment so as to satisfy one's needs and desires and the needs and desires of those affected by one's conduct. Therefore, truth is not what corresponds to facts, but that which satisfies one's conduct. Therefore, truth is not what corresponds to facts, but that which satisfies one's needs and desires and the needs and desires of others.

There is no absolute right, no self-evident or universally valid rules of conduct, in such a system. What is right or good is the harmonious satisfaction of one's needs and the needs of others affected by the moral agent's conduct. Moral rules of good or right are simply hypotheses which have been found to work in most cases and therefore offer helpful suggestions for the future.

Dewey's influence would infect the American educational system.

Immanentism or pragmaticism denies the reality of the transcendent. The spiritual and metaphysical are replaced by mechanical processes. There is no room for philosophy or theology in this world view, for philosophy explores natural revelation and theology explores divine revelation—both of which are ultimately explorations into the transcendent.

When everything is focused on the immanent all becomes solely pragmatic. Love and intimacy become limited to the sphere of a limited vision of sex as infatuation or self-infatuation. Contrition becomes equaled with a guilt complex, happiness with the Freudian concept of pleasure, moral values with superstition or convention.

Those who find pride in their immanentism fail to realize the implications and dangers that flow from such thinking.

Immanentism fuels a society that is bound up in conscupiscence and devoid of authentic personal and societal freedom. True freedom can only be experienced in the context of transcendence; freedom without transcendence is simply an illusion covering one's slavery to sin. Self-centeredness reigns. Even things done in the name of charity or other-centeredness are ultimately hidden or false for they are ultimately guided by self-interest, self-advantage.

When Christianity is infected with this affliction, the priest or minister becomes a preacher who preaches the gospel of the "here and now," the "gospel of wealth," the "gospel of health," the "gospel of political correctness," a gospel devoid of sacrifice, penance, contrition, mortification, a gospel devoid of the cross!

Logical positivism

Moral judgments do not state facts about the world but express emotional or psychological attitudes of the will. "Pleasant," "good," "truthful" and "right" are subjective and have no necessary connection to the world or to statements of facts. Thus disagreements are based on attitudes and not on facts: "If that is how you feel about it, there is no more to be said."

Amoralism

Our abandonment of critical thinking and the simple acceptance of the philosophies which have infected our thinking processes have given birth to amoralism. We have given life to an amoral mindset in our society which produces immorality. Blindness to moral values, an indifference to the questions of moral good and evil, and the blind acceptance that good and evil are simply conventions, superstitions, or taboos, has given a rebirth to paganism.

The modern world view, by attempting to make everything at the conscious or subconscious level neutral or as a mere physiological process has led to the elimination of the categories of good and evil.

This provides for a vision of the world that is based on material goods, earthly welfare, the cult of the body, and scientific progress.

Sadly to say this is viewed by many as liberation. In reality, this amoral view is nothing more than slavery to fallen nature's most base instincts, those instincts that make us no different than a high functioning lizard or monkey.

Idolatry

When one rebels against absolute truths, one becomes a prey to idols. One becomes a worshipper or disciple of opinion or mood. When one is a disciple of opinion or mood, when one's inclinations and tendencies are the source of the decision making process, then one is well on the road to debauchery and anarchy.

Dethronement of truth

Feelings, moods, slogans, propaganda, subjective perspectives have become the new standard for what is acceptable. Self-centeredness has replaced other-centeredness. Meanings of words are replaced by their emotional impact. One ideology that loses its popularity is replaced with another fresh ideology. Truth has been replaced with expediency, progress, pragmatism, with the "spirit of the age," with the "flow of thought" that is "in the air" and "up to date." Truth is seen as being in conformity with the culture of our time or with the national way of life practiced. Often truth is confused for anything that is practical or indispensible for our way of living. Many have replaced truth with historicism; that is, the thoughts or insights of one generation are true for that generation, but as one progresses in insights and knowledge truth changes for the new epoch or era. Many have replaced physiology for truth—the psychological reasons for belief as being equal to truth.

Two contradictory statements can co-exist as being equally good and equally valid. Something becomes true for one person and false for another. Brutal force replaces right; oppression and suggestive influences supersede conviction; fear supplants trust. Theism is replaced by practical atheism, practical secularism.

Blindness to value

Much of modern society has distorted the nature of authentic values. People only see value in purposes and ends, as opposed to seeing value in the directing of purposes and ends. The "ends justify the means" vision of life takes over in such a society.

Loss of reverence

Without reverence, which distinguishes humans from animals, there can be no true individual or societal knowledge. Without humility, a receptive character, self-abandonment, the acceptance that there is something greater than oneself, there can be no authentic fully human knowledge. Without reverence one enters into the sphere of the "survival of the fittest," and the "throwing away" of the unproductive.

How can one come to the knowledge of the mystery of love without reverence, to the mystery of life and children, to the mystery of the person, of responsibility, of friendship, and so forth. Is it any wonder that love has become equated solely with sex? Is it any wonder that sex has lost any sense of the mystery and is only viewed from a biological, psychological, scientific point of view? Is it any wonder that the world is becoming devoid of politeness and chivalry? Is it any wonder that we live in a society where vulgarity, lack of class, lack of shame and the lack of formalities is the norm? Is it any wonder that a crucifix in urine and a Madonna sculpted out of elephant dung is called art? Is it any wonder that divorce affirms a relationship that was based on infatuation rather than love? Is there any wonder that a person is thrown away through the use of the death penalty, abortion, and euthanasia, that violence is on the rise, that serial killers have become the movie stars of the 21st century? Is it any wonder that sex is not an act of unitive love but a mechanical act of contracepted self-pleasure and that children are not necessarily the product of the love between husband and wife but an act of artificial insemination, cloning, or inappropriate forms of genetic engineering?

Deification of Science

Since the sixteenth century the advances in the natural sciences have led to a conflict with philosophy and theology. The implication that developed was that somehow knowledge and faith were not reconcilable. The more that was discovered in the field of the natural sciences the less God was needed.

God became a concept used to fill the gaps of understanding (i.e., for Newton, God was the one who corrected the abnormalities in the orbits of the planets), but as science discovered plausible alternative explanations the need for God becomes less apparent.

The great paradox, however, of secularism is its divinization of science and the scientific method. The great irony and flaw of such an approach is that the very nature of science and the scientific method is based on first principles which can only be found in philosophy, ethics, metaphysics, and epistemology. The great irony is that relativism and secularism deny the very underpinnings of science itself.

Science, divorced from the divine, becomes a new religion. In such a world view truth itself, beauty itself, goodness itself, and even existence itself is beyond the scope of inquiry. One's purpose, meaning and eternal destiny are likewise beyond the scope of inquiry. These realities cannot be addressed by science alone for they go beyond the scope and limits of science and scientific inquiry.

Political Correctness and the dummying down of culture

When one fails to use one's freedom of thought effectively and rationally, then one ends up resorting to or compensating with forms of speech that are detached from thought and rationale. Wisdom and truth, arguments and proofs are displaced for slogans (i.e., the comforting sound or slogan of "pro-choice" replaces the reality behind the slogan, the killing of human life).

Political correctness has produced the dumb modern man, the man who views intelligence and progressive thought as an amalgamation of slogans. Truth is no longer the conforming of a statement to a reality; rather it is much more an expression of the "fashion of the time."

Political correctness has produced situation ethics—the replacement of abstract and absolute truth for "personal, subjective truth"—pseudo-personalism, the demythologization of the transcendent, the rationalization of mystery, the glorification of science, the laboratory view of the world, and the abortion of faith and the metaphysical.

Political correctness has continued the ongoing "sapping of the truth" in our society. Political correctness is at its very best "sloppy thinking."

Metaphysical indolence

Metaphysical indolence is a type of laziness of the mind that affects apprehension and that renders any real penetration of an object impossible. It prevents any collaboration with the meaning, nature, and essence of a thing. The deeper strata of existing things, or higher kinds of objects of knowledge, have been lost by this dummying down of the modern culture.

Indolence is a passive, dull, lazy way of remaining attached to that which one is accustomed to. Metaphysical indolence closes whole spheres of reality to the intellectual faculty.

Life is inevitably lived at the material level. It is a life whose primary concern is with that which is apprehended or presented in a definite manner within a definite field of vision. The sciences and the medical professions, such as modern psychology and psychiatry, for example, have so limited their view of the human person—to that of a material being, as opposed to a being consisting of body, soul, and spirit—that they have deeply harmed the scope of their effectiveness.

This materialistic, dummying down of the intellect, has led to the pursuit of biology without grasping the nature of life, psychology without grasping the nature of the person, sociology without understanding the true essence of community. Success over behavior and over good and evil, and the clinging to empiricism exclusively (seeing reality from the outside only) over eternal truth has led to a deafness, resentfulness, and a close-mindedness to anything beyond the superficial.

Deafness is the quintessential dimension responsible for the dummying down of our culture. A lack of will to listen, to let things speak and instruct us, to wonder, to penetrate, to thirst, to be open-minded are marks of this deafness. An inability for apprehending differentiation, mystery, and the endlessness of things is also a mark of this deafness.

Resentment or an unbending slavery to a distorted, ill-informed intellect and conscience, to a distorted fundamental attitude, an attitude that refuses to accept the objective, the autonomy of things, and the reality of the absolute are marks of a dumb society.

A distrust toward things, an inability to say yes or no, an inability to accept responsibility for a conclusion, a blindness to the self-evident, no matter how obvious or self-evident, an unbending slavery toward psychological predispositions, and unwillingness toward self-examination and self-reflection, and an unwillingness to be open-minded have contributed to the superficiality of the modern world.

We live in a society that seeks comfort, the easy way, and as little bodily and mental strain as possible. We live in a society that lacks virility and self-control and an inability for self-donation. We live in a society where the lack of discipline and self-mastery, which are necessary for love, are virtually non-existent. The thirst for truth, a desire to receive, a willingness to become empty, to keep silent, to listen to the universe, to be other-oriented has been lost.

What was lost!
Metaphysics

Metaphysics is a portion of philosophy dedicated to the most fundamental aspects of being and existence. It focuses on the nature of being itself (ontology), the nature of the human soul or life principle (rational psychology) and the reality and attributes of God (natural theology). Its emphasis is on objective truth and substantial reality.

Metaphysics is the study of being as being, the study of "beingness." Everything that exists, all being, the whole world or domain of reality, whether subjective or objective, possible or actual, abstract or concrete, immaterial or material, infinite or finite is the subject of metaphysical inquiry.

Metaphysics, unlike the fad of modern philosophy, does not abandon the study of the nature of the immaterial or transcendental. Metaphysics studies the immaterial being in the sense of that which is without matter (i.e., the immaterial life principle that distinguishes a dead body from a live body) or in the sense of abstract concepts (i.e., cause, quality, etc.).

Metaphysics is the study into the causes and principles of things. It seeks to inquire into the nature of *cause* and *principle* and to determine the meaning of the different kinds of causality.

Since metaphysics is ultimately the study of "beingness" it is concerned with that which is coextensive with being itself, namely unity, truth, goodness, beauty, and so forth.

The natural law

The natural law is based on the laws of nature and the meaning of the natural order of things as well as those innate instincts and emotions common to people and animals, such as the instinct for self-preservation.

Actions are "moral," "good," and "right," when they conform to the natural law, to that which is in conformity with the destined end of the human person, to that which is in conformity with a person's innate human nature and his or her manifold relationships.

Our actions are immoral when—acting in freedom and according to the guide and dictation of right reason—our actions are at variance with our nature and with the natural law.

Actions are good and right when they are fully human actions, and actions are immoral when they are less than fully human actions. Actions are good and right when they are in conformity with our human nature and the universal order of things.

Freedom, an informed or enlightened conscience—the exponent of the natural law—and the guide and dictation of right reason are at the heart of moral decisions according to the natural law; they are at the heart of perceiving the moral constitution of our nature.

Metaphysics and the natural law are at the core of understanding the reality of absolute truths, and therefore everything that flows from such absolute truths. When absolutes are lost, anarchy is just around the corner! Secularists deny the reality of metaphysical realities and therefore give birth to anarchy.

More Consequences!
Rupturing of motherhood, sexuality, and procreation

When a culture ruptures motherhood, sexuality, and procreation, it ruptures motherhood's point of reference. When the divinely-willed and nature-willed union of motherhood, sexuality, and procreation are artificially ruptured, then motherhood, sexuality, and procreation become devoid of authentic meaning and purpose.

The new point of reference becomes a personal, individualistic, self-centered, subjective, and self-evaluative vision of life. A distorted vision of pleasure—which is ultimately not pleasure at all—and the *libido* guides one's actions. Thus artificial insemination, contraception, abortion, cloning, the unmitigated hybridization of species, embryonic stem cell research, homosexuality, pre-marital sex, euthanasia, and so forth are all a matter of personal preference and cultural acceptance imposed by secular imperialism. When the culture's mores change, the personal preferences change as well.

Distorting the nature of woman

Who suffers most in the distortion of the nature of women? Women!

Many have attempted to mitigate the distinctions between men and women in order to pursue personal agendas. In this process, the "nature of woman" has been degraded and harmed. In the name of radical feminism, the feminine has been diminished.

The secular vision of equality between men and women is seen—in this distorted vision—as sameness, as interchangeableness. What is not realized is that the equality between the sexes implies diversity! Men and women are equal, but different. They are equal, but complimentary. Both are called to equal and eternal destinies, but differently.

Whether one is a male or female in our society is trivialized; we are just humans—as if sexuality was not rooted in anthropology. This trivialization makes the roles between males and females culturally and historically conditioned. Thus sex is not a part of a person's innate nature, but is an accidental function determined by culturally and historically determined roles (hence, the cultural acceptance of homosexuality, bisexuality, transexuality, premarital sex, open marriages, etc.).

Attempts to masculinize women in our secular society has come at a great price. It has contributed to the disintegration of the family, of the nature of motherhood and womanhood, and has damaged the nature of spousal life and love. This in turn has brought damage to our culture.

To trivialize the diversity of the sexes is to trivialize the nature of God, for we are created in his image and likeness, where there is a distinction, yet a oneness. Respecting biology and anthropology is therefore respecting God. To strip away the uniqueness of the sexes is to diminish the magnificence of God's creation.

Yet it is this failure to recognize the gift of diversity, uniqueness, and complementarity that has been the source of so much hurt for women in our culture and in the Church.

Distortion of the nature of man

What can be said of women above can in part be said of men. The nature of man has been distorted in the name of achievement above personality, production over content, production of impersonal goods above virtue, efficiency above all.

The distorted nature of man and woman continued

The nature of the person is diminished when the person views himself or herself as the owner of himself or herself.

The person in our secular society is no longer viewed ontologically. A person today is described more for what he (she) does than what he (she) is; that is, the person's essence has been lost to a world where achievement and efficiency are the only relevant factors. The person is a mere means of production of goods, an instrument to be used and discarded as a worn-out tool when the productive value of the person is no longer efficient. Death is the cost of doing business, a corporate loss.

Personality, character, and holiness are of little value as long as efficiency is maintained. Personality and character are of value to the extent that they foster achievement.

Workers have become extensions of machines, detached from their work, and therefore have lost not only the sense of what it means to be a person but have lost even in this product-oriented approach most of their meaning and purpose as people.

Mother Teresa of Calcutta's axiom has been completely lost in our modern society: "God only demands us to be faithful, not successful." Modern society has reversed this Christian axiom to be: "What is demanded of us is success, not faithfulness."

The outgrowth of not seeing a person in terms of his or her essence (that is, ontologically) is the loss of personhood in our society. This has led to the tragedies of abortion, eugenics, euthanasia, cloning, embryonic stem cell research, social disparities, prejudices, racism, and so on. The semi-deification of actors, athletes, corporate raiders, and even serial killers is the product of a production, achievement oriented society.

This secular vision has given existence to a society where the cult of personality is supreme. Mother Teresa's death had less an impact on the world than Princess Diana's. The advice of actors, athletes and businessmen is more important than that of philosophers, theologians, and academics. An actor or rock star with barely a high school education has more influence on the policies of a country than philosophers and intellectuals. Serial killers are now stars.

The cult of personality has led to the cult of the body. Hospitals have become repair shops, inspection facilities. Cosmetic surgery is no different than having one's teeth straightened with braces. The youthful beautiful body is what is important; one's personality or essence is secondary at best! Bodies and even decapitated heads are being kryogenetically frozen with the hope of immortality.

Work is no longer a work of God, an apostolate. Workaholicism is the norm and this workaholicism leads to the dummying down of the person. No longer can one afford to reflect, ponder, or pray on his or her essence as a human being.

The more one tries to make oneself into a god, the more one becomes an ape!

The media, secularism's evangelizing tool

The Catholic saint Elizabeth Anne Seton had a vision of a black box in every home, and from this black box the devil would come out and corrupt families and the world. Can anyone doubt that this black box is the television?

Television and the media are not a reflection of society as much as they are the means by which generations are socialized into a secular world view by secular imperialists.

In our quiet slumber we have allowed the media to be taken over by the secularists, with their pragmatic, positivistic, materialistic, egoistic, hedonistic, ultilitarianistic, evolutionistic, and relativistic, subjective vision of the world.

The religion of secularism has found the most powerful tool for its evangelization, the media, and in particular the television.

Anyone who has ever taken the most basic, elementary course in sociology is well aware that what we see, hear and experience has an impact on who we are and who we become. Today's culture is bombarded by television and the media. Is it any wonder that our society is being transformed so radically?

Violence

A plethora of sociological and psychological studies have shown that children and adults exposed to violent movies have a tendency to develop a heightened sense of aggressiveness. And for unstable children and adults this heightened sense of aggressiveness is exacerbated.

The American Psychological Association has pointed out through its research that people who view violence on television have a tendency to become less sensitive to the pain and suffering of others and are more apt to act out aggressively toward others. Children tend to become more aggressive as teenagers and tend to be more likely to engage in criminal acts as adults.

The secular obsession with control, personal preference, and with the survival of the fittest has corrupted many minds. We have generations of people who have been desensitized to violence. We have generations who have been desensitized to abortion, euthanasia, assisted suicide, and all forms of murder.

Cult of the body

The media is barraged by ads for all kinds of "beauty" products and "beautiful" people. The secular gods of self-absorption, self-preservation, self-pleasuring, and the viewing of oneself as a sex-object is the product of this cult of the body that is promoted in the media. Eating disorders, obsessive cosmetic surgeries, and a general unhappiness with one's self-image is the product of a media driven evolutionistic cult of superficial beauty. Physical appearance, and not one's innate nature or essence, is the dominant determinant of social value. The good-looking are the first to be hired, the first to make friends, and the ones who are most popular and preferred in our society.

And what about the so-called "ugly" and those who lose their beauty through age? They are relegated to a life of self-hatred, depression, and general malaise.

Cult of self-abuse

The cult of the body has its counterpoint in the media, the cult of self-abuse. If you can't be beautiful or if you can't see yourself as beautiful, then self-hatred needs an outlet. If you can't be beautiful then at the very least try to do what the beautiful people like to do on television.

Studies have shown that one of the prime causes of smoking, drinking and drug use in adolescents and adults is associated with media images.

The use of alcohol, cigarettes or cigars, and illicit drugs appears in seventy percent of prime time programs, over ninety percent of movies, and in half of all music videos (AAP New Release) .

A medicated society

The turn toward the self that is fostered through the evangelizing tool of secularism, the media, has made the self empty. With no god to fill the emptiness, drugs have become the tonic of choice. Statistics regarding drug usage—illicit and legal—in the United States are astonishing. It is estimated that one in five people between the age of 15 and 59 have abused drugs (NISA, 2001). With an ever increasing trend in mental health problems—nourished by the turn toward the self—drug usage and mental healthcare will continue to increase, reaching epidemic proportions.

A sex-obsessed world

When a person's essence or innate nature is of little value, then one becomes nothing more than an object like any other object—an object to be used. The average American will view nearly 15,000 sexually orientated images per year. Fifty-six percent of American television is loaded with sexual content. Family hour television averages eight or more sexual incidents a day. Daytime television has over 150 acts of sexual intercourse per day, with unmarried couples outnumbering married couples three to one (Pediatrics 107, Jan. 2001, 191-4).

Secularism's gospel of hedonism and self-pleasuring runs uninhibited. Modesty, purity, and inhibitions are thrown aside. We live in a society where sexual addiction and pornography are part of everyday life. This has led to a sexually permissive and perverted society. It is worth noting that the pornography industry is the largest industry in the world!

Din of vulgarity continued

Television promotes a society that is rude, loud, boastful, hostile, insulting, boorish, disrespectful, manipulative, and unkind in word and action. Disobedience, bullying, yelling, whining, rowdiness, throwing tantrums, and protesting traditional values are the norm.

What do we expect? A world without absolutes, without objective morality, without a respect for the natural law and the order of things, a world without a sense of the essence of the human person and human dignity cannot but be vulgar and cannot but promote vulgarity.

Dummying down of society revisited

As alluded to earlier, the lack of absolutes and the distrust for abstract realities and the transcendental in general has led to the largest population of "sloppy thinkers" the world has ever produced. In the midst of a plethora of information, much is absorbed consciously and subconsciously, but very little is reflected upon. There is very little time for reflection when one is obsessed with self-preservation, avoiding pain, self-pleasuring, survival, and emotional and psychological self-centeredness.

Studies have shown that viewing television hinders brain development. There are indications that television impacts the prefrontal cortex which is responsible for deliberate thought involved in planning and judgment. The bombardment of information and video stimulation seems to idle this part of the brain and stunt normal growth (cf. AAP News, May 1998, 2).

Not only have philosophical predispositions changed, one may even argue that the mind has changed—changed for the worse.

Subliminal seduction

Advertisers have used subliminal words and symbols in advertisements for years. They have used "buzz" words to grab readers' attentions, flashing images and quick messages. They have personified inanimate objects with qualities like wealth, fame, and success. They have manipulated what is cool and popular and what is nerdy. They have manipulated the minds of viewers with what kind of clothes to wear, what kind of car to drive, what kind of hairstyle is in, and so on.

If subliminal advertising were a simple fallacy then there would be no purpose for advertisements, would there? If subliminal advertising were a simple fallacy, why do we need bikini clad women selling cars, alcohol, lawnmowers, or even hamburgers?

Television and the media in general slowly seduce the viewer into accepting a particular vision of the world. And that vision is based on the religion of atheism, the religion of secular humanism. If you are not in line with this vision, you are just out of touch with things!

Whole generations have subliminally been seduced into worshipping at the foot of the gods of secularism—and sadly most have no idea whatsoever what has happened to them. They have no idea what they are swallowing. They are blind fish!

Loss of empathy

Empathy is essential for a moral society. Yet television has brought about a suffocation of empathy.

Families are spending less time working and playing together. Supper time is no longer seen as a time to gather around the table to share one's day. Conversation is becoming a lost art. Community activities, church functions, and civic functions were once the center of people's lives, now they are secondary at best.

We live in gated communities, with locked windows and doors, and alarmed homes. We too often don't speak to our neighbors or for that matter even know our neighbors.

Television has monopolized our emotions and need for others. It has manipulated our vision of reality into one that is superficial, self-oriented, self-preserving, and very lonely. If you doubt this, look at how many people walk their dogs every morning, noon, and evening. This is symptomatic of a loss of empathy in our society and a loss of connectedness. Animals have replaced people.

The god of materialism

Television promotes a life lived at its most primitive levels. As the cliché goes: "You only go around once in life, so you better get all you can get... The one with the most toys wins!"

Heroes are sexually promiscuous, violent, crude, and self-centered. They wear the latest in fashion and have the most expensive luxuries money can buy—thereby frustrating and embittering generations of individuals.

Careers are not intended for enhancing the common good or for bettering oneself, but are the means of acquiring the luxuries of self-infatuation and status. People are constantly selling their souls for their houses and luxury vehicles. They will bypass a good education for their children in favor of a status automobile.

Happiness is not about fulfilling one's innate moral calling. It is about living for the here and now, and to do so at whatever cost to others or society. As Blessed Mother Teresa would often say regarding the tragedy of abortion,

"It is a shame that a child must die so that you may live as you wish." Too many die throughout the world out of people's hedonistic needs. All in the name of the gods of secularism!

Too many think happiness is to be found in superficial relationships—where people are used up and thrown away when no longer needed. The ever frustrating unmet hunger for the perfect lover, the perfect sex partner, the perfect family, the perfect wife, the perfect husband, the perfect child, the perfect job, the perfect income, the perfect drug, the perfect body, and so forth, drowns the individual in a fantasy life of escapism, discontent, bitterness, and simmering anger.

The heroes of television are simply illusions of reality. They portray lives that are beyond attainment. And we wonder why our kids are shooting up their schools? And why serial killers become stars? And we wonder why marriages are breaking up? Who can live up to the impossible images that television portrays? Who is as beautiful or handsome, as popular, as smart, as witty, and as influential as these fake characters? Ah, all the lonely, discontented, and superficial people. We all know where they all come from!

The fostering of passivity

Television develops a passive society where the mores of the media are swallowed with little critical thinking. By virtue of the fact that a viewer is a passive receptor to an active purveyor, the television trains the viewer in passively accepting the superficial and often outright evil. The viewer has no impact on what is happening on the screen. He cannot act, debate, rebuke, protest, or challenge anything seen. Indecency, violence, and all forms of debauchery are paraded in front of the viewer unmitigated. A secular vision of the world is taught un-confronted. Thus, the more one becomes a passive viewer of secularism in front of the television screen the more one becomes a passive viewer of secularism in day to day life. One becomes trained to accept, to be meek, to be unconditionally accepting, to be uncritical, and to be unwilling to engage in intellectual speculation. One has been inculturated into secularism's sphere of "sloppy thinking."

What is normal becomes what is seen. The pleasures, habits, thoughts, language, and humor of a secular vision of the world becomes the norm. Blasphemy, obscenity, crudities, vulgarity and debauchery in all its forms become the norm.

By television's fostering of meek, submissive and passive minds, the television screen weakens and infects people's sense of right and wrong. It infects and even destroys consciences. The philosophy of majority opinion, consensus, preference, social pressure becomes the overriding principle of life by television's converting of souls to the religions of secularism. The gods of fame, power, drugs, and perversion reign supreme. Is it any wonder that a society that considers itself Christian is so unchristian?

Media in general

What has been said of television can be said to an equal or lesser degree about movies, theatrical performances, vice-ridden books or novels, tabloid journalism, journalism in general and magazines that flood the magazine stands. Wherever one looks or listens one experiences the religion of secularism, the religion of use and throw away, of self-pleasuring, of self-preservation. One experiences a religion where majority opinion and consensus is right, where the strongest, the so-called fittest are right, where individualistic psychological and emotional needs determine truth.

Is it any wonder that the quest for salvation and peace and happiness in God is replaced by the quest for sleek and sexy cars, sleek and sexy bodies, luxurious homes, and exotic status driven living? Is it any wonder that cloning is considered when one's immortality is based on the here and now, on this one and only life? Is it any wonder that society is tolerant, accepting and even endorsing of homosexual and bisexual acts, same-sex marriages, premarital sex, and so forth, and yet completely intolerant to a Christian vision of life? Is it any wonder that people are more concerned with saving seals or not wearing fur nor eating meat than they are about preventing the abortion of children? You can tear a child to pieces but do not dare lay a hand on a seal! You will pay a fine and even receive prison time for killing a manatee but will go away free and clear after aborting an infant in the womb! And where is the outrage, the indignation? It has been pacified under unconditional acceptance and false tolerance.

Secularism, in conclusion

What has the growth of secularism done for the world? Has it brought the world and individuals closer to what the atheists or secularists thought it would? Has secularism made things better?

Any system of belief such as agnosticism or secularism that does not recognize the transcendent is bound to disintegrate into debauchery. The great failure of secularism is found in its naïve anthropology and its refusal to recognize the self-evident and what is often referred to as common sense. Secularism fails to recognize one's innate inclination toward the dysfunctional and one's innate inability to control one's future and one's surroundings. It fails to recognize the human person's natural inclination toward self-centeredness, toward being self-oriented as opposed to being other-oriented. It fails to recognize the demands of or even the existence of the natural law and the existence of absolutes such as truth.

The rise of secularism has not brought about its hoped for freedom or personal autonomy. It has brought about the enslavement of a society to its passions, desires, and dysfunctions. It has brought about a revival of paganism, barbarism, and anarchy.

Whereas belief in God was always a restraining force to the innate evil inclinations of the human person, the abandoning of God in our modern culture has brought about the moanings of its death. Secularism is birthing an unrestrained culture that inevitably will become a den of thieves, murderers, gluttons, and perverts. The ever increasing waves of nominalism, pragmaticism, hedonism, positivism, and relativism will continue to poison the world.

Without absolutes as a levee of protection, a society flushes out whatever good it has. When everything is based on usefulness, self-preservation, hedonism, consensus, the survival of the fittest, then a culture disintegrates. The debauchery of yesteryear becomes the acceptable practice of today. In the past, contraception, abortion, euthanasia, homosexuality and divorce were considered unmentionables. Today these unmentionables have become perfectly acceptable. They have become politically correct.

Secularism, by nature, leads to the death of a culture. It leads to a pagan, barbaric, anarchistic culture. Just as the lives of the world's most famous atheists, secularists, have ended up in anger, bitterness, despair, and moral and mental collapse, so too do the cultures that they sought to build: Nietzsche, the god of atheism, became a bumbling, broken down wreck.

The belief in that which transcends oneself is the only factor that can keep a person and a culture from disintegrating—for the belief in that which transcends oneself is the only thing that can keep one answerable to something other than oneself.

X
MORAL DOCTRINES UNDER ATTACK

The Ten Commandments and their implications for Catholics (cf. Exodus 20:2-17)

1. The first commandment forbids acts of superstition, divination, magic, and all forms of sacrilege. It forbids acts of idolatry such as the worship of money, power, fame, and all sorts of "worldly" accomplishments. It forbids atheism and agnosticism, for they are nothing other than the hidden or subconscious worship of self.

2. The second commandment demands a respect for the sacredness of the Lord's name. Acts of blasphemy, the taking of false oaths, and acts of perjury are strictly forbidden.

3. The third commandment is a summons to keep the Lord's Day a holy day. It demands the faithful attendance of Sunday Mass, and an attitude of profound worship. It is a time to spend with God and to abstain from any work that distracts from authentically consecrating Sunday as a precious day of love of God and love of neighbor. One seeks comfort, but one also seeks to be challenged to grow.

4. The fourth commandment demands the authentic honoring of father and mother. This means obedience, respect, gratitude, and the repaying of love for love.

5. The fifth commandment is an affirmation of the dignity of life, of not murdering. Unjust war, direct abortions, the use of contraceptives, suicide, and intentional euthanasia are all forbidden by this commandment.

6. The sixth commandment is a command that demands fidelity. Any act which is contrary to the dignity of chastity, such as fornication, adultery, polygamy, open or free marriages, divorce, homosexual and bisexual acts, masturbation, and pornography are forbidden. The sixth commandment is a call to authentic sexual integration.

7. The seventh commandment is a prohibition against stealing. It is characteristic of a lack of charity and injustice. Often stealing is done in subtle ways: For example, on the part of employers in a business a violation of the seventh commandment is often exemplified by the mistreatment of workers through unfair wages,

lack of health benefits, and lack of retirement benefits. On the part of the employee this injustice and lack of charity is often seen in acts of laziness and all forms of lack of effort in the work environment.

8. The eighth commandment is a prohibition against bearing false witness against one's neighbor. Lying, duplicity, hypocrisy, dissimulation (that is, hiding under a false appearance), betrayal of confidences, calumny (character assassination), slander, and so forth are all acts contrary to the dignity of persons.

9. The ninth commandment is a prohibition against coveting one's neighbor's wife. This commandment calls one to live a life of decency and modesty. It is a call for purity of heart, intention, and vision.

10. The tenth commandment is a call to avoid coveting another's goods. It is a call to avoid avarice, envy, and all immoderate desires. It is a call to desire a detachment to all that is contrary to the glory and honor of God. One is called to desire God above all.

Fulfilling the Commandments

[A lawyer asked Jesus:] "Teacher, which is the greatest commandment in the law?" And he said to him, "You shall love the Lord your God with all your heart, with all your soul, and with all your mind. This is the great and first commandment. And a second is like it. You shall love your neighbor as yourself. On these two commandments depend all the law and the prophets" (Matthew 22:37-40, RSV).

To authentically love is to fulfill and grasp the true intentions of the commandments. The "culture of life" and the roadmap to light, peace, and happiness are based on the fulfillment of these commandments.

The natural and moral law, essential dimensions to moral theology

The Bible alone approach can often lead individuals to miss another important aspect of revelation, natural revelation. Natural revelation is based on the natural law and the laws of nature that God created. By being aware of the inner principle, called conscience, one can know right from wrong. By observing the order and structure of God's creation one can likewise know the right from the wrong in terms of our interactions with all of creation and God. When natural revelation is corrected by divine revelation in Scripture and Tradition then one can come to a knowledge of God by human reason and have a greater understanding of the world. Faith seeks understanding and the more we understand the more our faith is nourished.

Thus, a sin is not only that which is condemned in the Bible or Sacred

Tradition but it is also condemned by the natural or moral law. What is contrary to the natural and moral law is a sin. In Genesis 1:1-2:4 a perfectly ordered, harmonious world is created. This is followed by the Fall, the "original sin" (Genesis 3) where this harmony and order are destroyed. As we will see in the case of homosexual acts and the other moral sins, they are sins that are condemned by God in the Scriptures, Tradition, and by the very nature of God's creation.

The Ten Commandments above are not arbitrary laws or rules of conduct; rather, they are the expression of that God-given reality which is at the core of every human being guiding him or her into the ways of righteousness or depravity.

Why is homosexual activity (and same-sex marriages) contrary to the Word of God?

The Catholic Church basing "itself on Sacred Scripture, which presents homosexual acts as acts of grave depravity, [and] tradition has always declared that homosexual acts are intrinsically disordered" (CCC 2357; CDF, *Persona Humana*, 8). "They are contrary to the natural law. They close the sexual act to the gift of life. They do not proceed from genuine affective and sexual complementarity. Under no circumstances can they be approved" (CCC 2357).

Scripture is clear. The story of Sodom and Gomorrah in Genesis 19:1-14, while often argued as an account of inhospitality, is an account of the evil of homosexual activity; otherwise, why would all generations call those who perform homosexual acts sodomites? Furthermore, has there ever been an account of God destroying an entire city with fire and brimstone for its failure of showing hospitality?

Leviticus 18:22 states: "You shall not lie with a man as with a woman; such a thing is an abomination." Leviticus 20:13 states: "If a man lies with a male as with a woman, both of them shall be put to death for their abominable deed; they have forfeited their lives."

Now some like to argue that there are many things forbidden by the Hebrew Scriptures which are no longer held by Christians. There are those laws which Jesus specifically addressed as in the case of what to do with a person caught in adultery (Jn. 8:3f) or in the case of those suffering from leprosy (Lk. 5:13). There is the example of the apostles eliminating the law of circumcision (Acts 15). And there is the making of what was once "unclean" clean in Peter's revelation (Acts 10:9-33).

The point is that unless Jesus and his Church specifically clarified and overturned certain Hebraic laws, the laws were to remain. Leviticus forbids sex with your mother (18:7), with your sister (18:9), and with your aunt (18:14). It forbids bestiality (18:23) and orgies (18:23). I don't think that those who favor the overturning of the Hebraic laws are in favor of practicing these evils.

But let us look at the New Testament writings written after the death and resurrection of Christ, when the Spirit of truth (Jn. 15:26; 16:13) was sent to the Christian community. Furthermore, let us never forget the promises of Christ, the promise that the gates of hell would not prevail against his Church (Mt. 16:18f; Jn. 16:13; 28:20; 1 Tim. 3:15) and the promise that he would be with his Church till the end of time (Jn. 20:29).

Let us remember that the letters to Timothy, to the Romans, and to the Corinthians in the Bible were written by Christ's greatest theologian. Paul, who lived after the resurrection of Jesus! If it wasn't for Paul, we would know very little about Christ, his Church, and Christianity in general!

In 1 Corinthians 6:9-10 we read: "Do not be deceived; neither fornicators nor idolaters nor adulterers nor boy prostitutes nor sodomites...will inherit the kingdom of God."

In Romans 1:26-27 the Scriptures declare: "Their females exchanged natural relations for unnatural, and the males likewise gave up natural relations with females and burned with lust for one another. Males did shameful things with males and thus received in their own persons the due penalty for their perversity."

In 1 Timothy 1:10-11 we read: The "law is meant not for a righteous person but for the lawless and unruly, the godless and sinful, the unholy and profane, those who killed their fathers and mothers, murderers, the unchaste, sodomites, kidnappers, liars, perjurers, and whatever else is opposed to sound teaching, according to the glorious gospel of the blessed God, with which I have been entrusted."

But it is not simply individual quotes that condemn homosexual acts, the very theology of the Old and New Testaments condemn it. The underlying theology of God's love for his people in the Old and New Testament is based on the complementarity of the sexes and on the natural law which underlies this complimentarity (Genesis 1 illustrates how the complimentarity of the sexes reflects God's inner unity). Men and women are physically and psychologically different, and it is in this distinction that the complementarity between a man and a woman make the possibility of two becoming one through total self-donation (cf. Gen. 2; Mt. 19:3-6; Mk. 10:6-9). The theology of Genesis and the entire Pentateuch as well as the theology of the Wisdom and Prophetic books of the Bible are all based on the underlying theology of the love of God for his people in the form of the love of a man for a woman in their distinct natures. In fact, there is no way of understanding the Scriptures without understanding the relationship between the sexes!

Tradition is clear. Some sixteen centuries before the birth of most Protestant denominations, Christians believed that homosexual acts were contrary to the will of God.

In the *Didache, The Teaching of the Twelve Apostles*, written anywhere from 65 AD to 120 AD, we are told to "not be sexually perverted by committing sodomy" (cf. 4). In Polycarp's *Letter to the Philippians*, the disciple of the apostle John, Polycarp states: "Sodomites shall not inherit the Kingdom of God." And in Barnabas, often attributed as the same Barnabas who was the companion of Paul, we read: "Thou shall not commit sodomy" (n. 19).

Never, ever has the approval of homosexual acts been accepted in Church history prior to the twentieth century!

Philosophy likewise is clear. To put it bluntly a male's genitals were not created for another male, and a male's sexual organ certainly has no place in any male body! The male and female organs are complementary, just as the psychological distinctions between males and females are complementary. The homosexual act is a sex act which is contrary to the act's purpose: It is an act completely closed off to physical and spiritual life.

Because of the nature of males and females, the sexual act is unitive and procreative. Homosexual acts are neither unitive nor procreative, and thus are a direct attack on the dignity and the sanctity of the sexual act. Homosexual acts are masturbatory acts.

In pagan societies homosexual activity was common and even practiced as part of many cults. It was so common that students that take college courses in Greek and Latin Classics are often shocked by the open discussion of homosexual activity in these cultures.

During the period of the early Church, the distinction between homosexual activity and homosexual orientation was not made, being that it was so closely associated with paganism. It is only with the Church's correct interpretation, guided by the Holy Spirit, that the distinction between orientation and activity was made.

The Church makes it clear that a person's orientation is not sinful. As the *Catechism* states: Homosexuals "must be accepted with respect, compassion, and sensitivity. Every sign of unjust discrimination in their regard should be avoided. These persons are called to fulfill God's will in their lives and, if they are Christians, to unite to the sacrifice of the Lord's Cross the difficulties they may encounter from their condition" (CCC 2358). The *Catechism* goes on to say: "Homosexual persons are called to chastity. By virtue of self-mastery that teach them inner freedom, at times by the support of disinterested friendship, by prayer and sacramental grace, they can and should gradually and resolutely approach Christian perfection" (2359).

What has been said about homosexual acts are what make same-sex marriages contrary to God's Word and Will.

No such thing as pro-choice Catholics!

It is heretical to be a pro-abortion Christian. One cannot be a Catholic in good standing by maintaining a pro-abortion stance, and one cannot vote for a pro-abortion politician in good conscience!   Direct abortion is intrinsically evil!

The Scriptures

The Scriptures are clear regarding the sanctity of life from conception to natural death: In Genesis 25: 22-24 we read: "The children in Rebekah's womb jostled each other so much that she exclaimed, 'If this is to be so, what good will it do me!' She went to consult the Lord, and he answered her: 'Two nations are in your womb….'" In Jeremiah 1:5 we read: "Before I formed you in the womb I knew you, before you were born I dedicated you, a prophet to the nations I appointed you." In Isaiah we read: "Thus says the Lord who made you, who formed you from the womb: Fear not, O Jacob, my servant whom I have chosen" (v. 2 and v. 24). In Isaiah 49:2 we read: "The Lord called me from birth, from my mother's womb he gave me my name." In Job 10:8, 11 we read: "Your hands have formed me and fashioned me; with skin and flesh you clothed me, with bones and sinews you knit me together." And in Job 31:15 we read: "Did not he who made me in the womb make him? Did not the same One fashion us before our birth." In Psalm 139:13-16 we read: "You formed my inmost being; you knit me in my mother's womb. I praise you, so wonderfully you made me, wonderful are your works! My very self you knew; my bones were not hidden from you, when I was being made in secret, fashioned as in the depths of the earth. Your eyes foresaw my actions; in your book all are written down; my days were shaped, before one came to be." In Ecclesiastes 11:5 we read: "Just as you know not how the breath of life fashions the human frame in the mother's womb, so you know not the work of God which he is accomplishing in the universe." In Luke 1:41-44 we read, "When Elizabeth heard Mary's greeting, the infant leaped in her womb, and Elizabeth, filled with the Holy Spirit, cried out in a loud voice and said, 'Most blessed are you among women, and blessed is the fruit of your womb. And how does this happen to me that the mother of my Lord should come to me? For at the moment the sound of your greeting reached my ears, the infant in my womb leaped for joy.'" And in Luke 1:36 we read: "Behold, Elizabeth, has conceived a son in her old age, and this is the sixth month for her…"

Finally, in Revelation 9:21f we read: "Nor did they repent of their murders, their magic potions, their unchastity…" The phrase "magic potions" is from the Greek word *pharmakeia*, which means, in this context, an abortion causing agent.

Other quotes worth reviewing: Genesis 16:2-4; 19:36-38; 21:1-18; 38; 50: 20; Exodus 21:22-25; Leviticus 19:14; Numbers 35:22-34; Deuteronomy 27:25; Jeremiah 7:6; 22:17; Isaiah 45:9-12; Psalm 94:9; 106:37-38; Proverbs 6:16-19; Ruth 4:18-22; Matthew 1:3; 18:10-14; Luke 3:33; 17:2; John 9:1-3; Acts 17:25-29; Romans 8:28.

How can anyone understanding the Scriptures ever ponder the possibility of abortion!

Early Church Writings

In the *Didache* (ca. 65) the *Teaching of the Twelve Apostles*, we read: "You shall not kill an unborn child or murder a newborn infant" (II, 2). In Barnabas' *Epistle II* (ca. 70) we read: "You shall love your neighbor more than your own life. You shall not slay the child by abortion." In Tertullian's *Apologetics* (ca. 177) we read: "For us murder is once and for all forbidden; so even the child in the womb, while the mother's blood is still being drawn on to form the human being, it is not lawful for us to destroy. To forbid birth is only quicker murder. He is a man, who is to be a man; the fruit is always present in the seed" (197). In Athenagoras' *Legatio pro Christianis*, (ca. 177) we read: "Those who use drugs to bring about an abortion commit murder and will have to give an account to God for their abortion." In Minucius Felix's *Octavius* (ca. 200) we read: "There are women, who, by the use of medicinal potions, destroy the unborn life in their wombs, and murder the child before they bring it forth. These practices undoubtedly are derived from a custom established by your gods; Saturn, though he did not expose his sons, certainly devoured them." In Clement of Alexandria's *Christ the Educator II* (ca. 150) we read: "If we would not kill off the human race born and developing according to God's plan, then our whole lives would be lived according to nature. Women who make use of some sort of deadly abortion drug kill not only the embryo but, together with it, all human kindness." In Augustine's *De Nuptius et Concupiscus* (354-430) we read: "Sometimes this lustful cruelty or cruel lust goes so far as to seek to procure baneful sterility, and if this fails the fetus conceived in the womb is in one way or another smothered or evacuated, in the desire to destroy the offspring before it has life, or if it already lives in the womb, to kill it before it is born." In Jerome's *Letter to Eustochium* (ca. 340-420) we read: "Some unmarried women, when they are with child through sin, practice abortion by the use of drugs. Frequently they kill themselves and are brought before the ruler of the lower world guilty of three crimes; suicide, adultery against Christ, and murder of an unborn child." In Basil the Great's *First Canonical Letter* (ca. 329-379) we read: "The hairsplitting difference between formed and unformed makes no difference to us. Whoever deliberately commits abortion is subject to the penalty for homicide." We could go on and on.

To call oneself a Catholic and pro-abortion or pro-choice is to promote an outrageous lie. To be indifferent or to vote for pro-choice or pro-abortion candidates is to betray one's Catholic faith.

One may argue that there are many issues in life and abortion is just one of the many injustices in life. The answer to such a statement is quite clear. If you could wipe out all the hunger in the entire world, you could not justify one abortion! If you could wipe out all the homeless and the disenfranchised in the entire world, you could never justify one abortion! If you could eliminate all illness and every other aliment in the world, you could never justify one abortion! An evil means does not justify what one may perceive to be a good ends!

Let us never forget the words of Caiaphas: "It is better for one man to die than for a whole nation to perish" (Jn 11:50). That one person was Jesus.

We were created in the "image and likeness" of God (Gen. 1:27). We are the "body of Christ" (1 Cor. 12:12f; Rom. 12:5; Eph. 1:22f) and the "Temple of God" (1 Cor. 3:9-10, 15-16). Anyone who aborts a child is aborting the "image and likeness" of God, the "body of Christ," the "temple of God." They are aborting God.

They are committing an act of sacrilege, for we are "not our own" (cf. 1 Cor. 6:19-20)—our bodies belong to God.

Why are contraceptives evil and why is Natural Family Planning holy?

> *Onan knew that the descendants would not be counted as his; so whenever he had relations with his brother's widow, he wasted his seed on the ground, to avoid contributing offspring for his deceased brother. What he did greatly offended the Lord, and the Lord took his life.*
>
> *Genesis 38:9-10*

As Genesis 38:9-10 illustrates, the use of contraceptives is viewed in the eyes of God as an evil. The Scriptures call couples to be "fruitful and multiply" (cf. Gen. 1:28). Couples are called to recognize that children are a gift from God to be cherished (Ps. 127:3-5; also 1 Chr. 25:4-5; 26:4-5). Sterility or childlessness was often viewed as a punishment—even perceived as a lack of God's blessing (Hos. 9:10-17; Ex. 23:25-26; Dt. 7:13-14; Lv. 21:17-20; Dt. 23:1; Dt. 25:11-12, etc.) In Revelation 9:21 the Greek word *pharmakeia* for "magic potions" was traditionally used to describe the evil of abortion causing "potions"—in other words contraceptives and abortifacients.

The Scriptures and the theology of the body as portrayed in the Bible make it quite clear that contraceptives are intrinsically evil. As *Familiaris Consortio*, 32, explains regarding the evil of contraceptives and the contraceptive attitude (as derived from the Scriptural understanding of the moral law):

> *The innate language that expresses the total reciprocal self-giving of husband and wife is overlaid, through contraception, by an objectively contradictory language, namely, that of not giving oneself totally to the other. This leads not only to a positive refusal to be open to life but also to a falsification of the inner truth of conjugal love, which is called upon to give itself in personal*

totality....The difference, both anthropological and moral, between contraception and recourse to the rhythm of the cycle...involves in the final analysis two irreconcilable concepts of the human person and of human sexuality.

How can two become "one" through the use of contraceptives (cf. Mk. 10:6-9; Mt. 19:3-6)?

Hormonal Methods

If we were to ask most couples about the negative side effects associated with the use of the pill, most couples would have a general idea regarding these effects, either through information obtained from their doctors or from pharmacists. They may not be aware of the fifty-two side effects associated with the use of the pill, but they more than likely would be aware of the most talked about side effects such as strokes, heart attacks, and blood clots.

If, however, we were to ask most couples about the method in which the pill works in preventing the birth of children, there would be a tremendous amount of ignorance.

There are two major types of pills that are being used in preventing the birth of children: those that contain a combination of estrogen and progestogen and those that contain only progestogen. Both of these types of pills prevent the birth of children either through preventing ovulation or preventing the effective migration of sperm in the uterus, or by preventing implantation. In the Physicians' Desk Reference the combination pills are described as operating in the following manner: "Combination oral contraceptives act by the suppression of gonadotropins. Although the primary mechanism of this action is the inhibition of ovulation, other alterations include changes in the cervical mucus (which increase the difficulty of sperm entry into the uterus) and the endometrium (which reduce the likelihood of implantation)."

In terms of the progestogen-only pill, the Physicians' Desk Reference states: "[Progestogen-only pills] alter cervical mucus, exert a progestational effect on the endometrium, interfering with implantation, and in some patients, suppress ovulation."

Therefore, the pill (whether the combination pill or the progestogen-only pill) has the potential for being an abortifacient—an abortion-causing agent. When conception takes place, a human being, a human embryo, is present. The pill at this point, because it weakens the lining of the uterus, prevents this human being, this human embryo, from being implanted in the womb of the mother.

This is a silent abortion. As the Church teaches in its documents, and in particular in the 1994 American document *Ethical and Religious Directives for Catholic Health Care Services* (n. 45), "every procedure whose sole immediate effect is the termination of pregnancy before viability is an abortion, which, in its moral context, includes the interval between conception and implantation of the embryo."

What is said of the "pill" can be said, with slight variations, of all the other hormonal methods of contraception including Norplant, Depo-Provera, RU-486 and Ovral.

Similar abortifacient effects are also apparent in the use of intrauterine devices such as the Lippes Loop and the Copper-T 380A.

How many silent victims are being lost because of the unknowing actions of couples? Who is at fault for their ignorance?

What about the non-hormonal methods?

Sterilization as a method of contraception is forbidden by the Church and the Scriptures (cf. Gen. 1:28; Hos. 9:10-17; Ex. 23:25-26; Dt. 7:13-14; Lv. 21:17-20; Dt. 23:1; Dt. 25:11-12, etc). As the *Religious Directives for Catholic Health Services* states (n. 53): *"Direct sterilization of either men or women, whether permanent or temporary, is not permitted...when its sole immediate effect is to prevent conception."*

Sterilization and barrier methods such as condoms, diaphragms, cervical caps, vaginal pouches, spermicidal sponges, suppositories, foams and jellies are prohibited because they damage the unitive and generative aspect of the sexual act. They are methods of birth regulation that are contrary to the natural order.

The above methods suppress total self-giving. They are acts that are self-centered rather than spouse-centered. They view the spouse as an object to be used. They are essentially masturbatory acts as opposed to acts of love. Is it any wonder that those who use contraceptives have an over 50% rate of divorce?

Contraceptives prevent sexual acts from being holy acts. They prevent any authentic bond between husband and wife and they prevent the act of love from being procreative—for authentic love is creative by nature. We often refer to the love of the Father and Son as the gift of the Holy Spirit. Love is creative by nature.

What is Natural Family Planning and how does it differ from contraceptives?

The old fashion "calendar-rhythm" method, which was highly inaccurate and inadequate, is no longer the means used for natural family planning. Today the methods for determining a woman's fertile period have become more sophisticated and accurate. Some prefer the use of the Ovulation-Billings method, others prefer the Sympto-Thermal method.

Those who practice one of these methods of natural family planning have a 4% divorce rate. Those who use contraceptives have a 50% divorce rate. The reasons for NFP's low divorce rate are simple:

NFP methods are natural. That is, they do not hinder the natural functioning of the body but observe and respect the natural cycle of fertility and infertility.

These methods respect the bodies of the spouses, encourage tenderness, and foster the necessary freedom that is at the base of authentic self-giving love.

In the practice of living out the natural methods one is engaging in a love which is expressed by the husband in saying, "I give you everything I am without doubt, without reservation, fully and completely," and the wife in turn says to her husband, "I give you my very self, completely, fully, without doubt, and without reservation." It is only in this grace-filled experience that the Gospel call of two becoming one can be fulfilled (cf. Mark 10:6-9).

Human life and the duty of transmitting it in cooperation with God is a spiritual gift that is not limited to this life's horizons, but has its true evaluation and full significance in reference to one's eternal destiny.

Those who practice natural family planning, as opposed to artificial contraception, make the sex act a spiritual act, a unitive, bonding, and creative act.

Artificial contraceptives—99% effective—50% divorce rate!
Natural family planning (NFP)—99% effective—4% divorce rate!

Why do Catholics believe in legitimate wars?

There is such a thing as legitimate war. The Old Testament is filled with stories of God's people fighting to do God's will: Moses against Egypt (Ex. 14), Joshua against Jericho (Jos. 6), the family of Mattathias against the Greeks (1&2 Maccabees).

The Lord asked Cain, 'Where is your brother Abel?' He answered, 'I do not know. Am I my brother's keeper?'

Genesis 4:9

It is true that as Christians we are called to "turn the other cheek" (Mt. 5:39); But as the quote from Genesis above points out, we are also "our brother's keeper" (Gen. 4:9). It is when we are called to be our brother's keeper that we can legitimately—as a last resort—turn to war. It is for this reason that Jesus overturned the tables of the "money changers" and chased them out of the temple with a whip (cf. Jn. 2:13-16).

Therefore, in a spirit of prudence, the Church affirms the legitimate right to self-defense and war: "The legitimate defense of persons and societies is not an exception to the prohibition against the murder of the innocent that constitutes intentional killing" (CCC 2263).

"Legitimate defense can be not only a right but a grave duty for someone responsible for another's life, the common good of the family or the state" (CCC 2265). "Governments cannot be denied the right of lawful self-defense, once all peace efforts have failed" (GS 79,4). In fact, governments are often called to war for the betterment and the good of the world.

Recourse to war is permissible when the following conditions are met (2309; 2313-2314; ST II-II, 64, 7).

1. The cause must be just.
2. All means of avoiding war or ending aggression must be seen to be "impractical and ineffective."
3. The "damage inflicted by an aggressor on the nation or community of nations must be lasting, grave, and certain."
4. There must be an adequate prospect for success in putting an end to the aggression or evil.
5. The use of weaponry must be used with prudence. They must not "produce evils and disorders graver than the evil to be eliminated."
6. Every act of self-defense or war that is aimed at the indiscriminate destruction of whole cities is prohibited. Non-combatants must never be targeted.

Acts of terrorism remind us of the challenge of peace that we as Catholics are faced with. Hostilities, excessive economic inequalities, contempt and distrust for persons, and unbending ideologies are all part of the injustices that ferment war (cf. Ex. 20:2-7). What is needed is a spiritual renewal throughout the world, a renewal that fosters solidarity and a sense of universal cooperation among nations. All nations are called to a spirit of brotherhood and a desire for a universal common good. Social structures, attitudes, and hearts must change (GS 83-90). Unless we take up this challenge for peace, the world will inevitably enter a new dark age. Recent events have pointed to this sad reality.

When the Commandments are ignored (cf. Ex. 20:2-17), when the love of neighbor and the love of God is ignored (cf. Mt. 22:37-40), when the golden rule of do unto others as you would like done unto you is ignored (cf. 25:31f), then wars become inevitable.

God created a harmonious, orderly world (cf. Genesis 1&2). Sin distorts this order and harmony.

The death penalty revisited

The logic behind legitimate wars is very important to understanding the theology regarding the death penalty.

Punishment for criminal offenses has traditionally emphasized the importance of justice, retribution, deterrence and the protection of the moral and structural fiber of society. It is in this way that the death penalty was used in the Old Testament (cf. Gn. 9:6; Ex. 21:16, 22f, 22:18; Lv. 20:10-15, 27; 24:16-17; Dt. 17:12; 21:9, etc.). The key principle in regard to the death penalty has always been the protection of society—either the physical or moral protection of society.

As Christians we are called to turn the other cheek (Mt. 5:39) when we can, but we are called to be our "brother's keeper" (Gen. 4:9) when turning the cheek is ineffective. It is for this reason that Jesus overturned the tables of the "money changers" and chased them out of the temple with a whip (cf. Jn. 2:13-16).

In describing the Church's position on the death penalty, the *Catechism of the Catholic Church* explains: "If nonlethal means are sufficient to defend and protect people's safety from the aggressor, authority will limit itself to such means, as these are more in keeping with the concrete conditions of the common good and more in conformity with the dignity of the human person" (CCC 2267). In other words, if the key principles behind the Old Testament understanding of justice are met without the need to resort to the death penalty, then that or those means should be used.

Many people who read this Catechism passage often scratch their heads while saying: "How can this be? Isn't this the Church that has affirmed and often promoted the death penalty for centuries? What is going on?"

At first glance there may appear to be an inconsistency in the Church's current teaching on the death penalty, but in reality the Church's teaching has remained absolutely consistent.

The change in the Church's position is not due to a change in its theology as much as to developments in the ways of protecting and defending the common good of society. Once again, if the key principles behind the Old Testament understanding of justice are met without the need to resort to the death penalty, then that or those means should be used.

Prior to the nineteenth century, violently dangerous criminals were dealt with by means of execution or exile (which was essentially another form of capital punishment due to the atrociously harsh conditions associated with it).

The infrastructure of society prior to the nineteenth century was incapable of dealing with long-term incarceration; hence, those who posed a serious threat to society, such as the criminally insane, needed to be taken out of society for the protection of the common good, and the only means available, for all practical purposes, during this period in history was the death penalty (Ives, *A History of Penal Methods*). It is for this reason that the Bible is replete with examples of the death penalty.

By the late nineteenth century and early twentieth century, however, developments in the structure and organization of society as well as enlightened thought led to the possibility of incarcerating individuals for life, thereby eliminating the moral justification for the death penalty. As Pope John Paul II explained in *Evangelium Vitae*: "Today...as a result of steady improvements in the organization of the penal system [the justification for the death penalty is] practically non-existent."

Justice without mercy is cruelty. Christian justice demands that we be protected from violent criminals, and Christian mercy demands that we forgive the unforgivable and hope for the hopeless. As long as there is life, there is the possibility for repentance and conversion (Lk. 23:39-43). There is always hope. Death extinguishes hope and any possibility of conversion. If Jesus would not pull the switch or inject a person with heart stopping chemicals, why should we? Let society imprison the dangerously uncontrollable for the remainder of their lives, and let people of faith pray for their conversion. Let us remember that "whoever brings back a sinner from the error of his way will save his soul from death and will cover a multitude of sins," and let us also remember that there is "more joy in heaven over one sinner who repents than over ninety-nine righteous persons who need no repentance" (Jms. 5:20; 5:7). And finally let us remember the thief on the cross next to Jesus who obtained eternal salvation at the very end of his life (Lk. 23:43).

In ancient times the death penalty was perfectly acceptable for the protection of the good of society—hence its use in the Old and New Testament period. In a modern, civilized society, the death penalty has no place.

Euthanasia

I have had lots of patients who wanted to commit suicide, but you don't help them do it. You learn why patients don't want to live anymore. If they're in pain, you give them more or better medication. If they have trouble with their families, you help them get the problem solved.

Elizabeth Kubler-Ross

Elizabeth Kubler-Ross was a world-renowned medical doctor and psychiatrist. She did much research and wrote several books and articles in the area of death and dying. In her research, she found that people who face death often experience episodes of denial, anger, bargaining with God, and depression. Most importantly, she pointed out that if a patient was lovingly cared for, the patient's last moments would be ones filled with acceptance and even hope.

Direct euthanasia consists in the murdering of the handicapped, the ill, and the dying—with or without their consent and knowledge—and is thus morally unacceptable (CCC 2277). In the definition used by the Congregation for the Doctrine of the Faith in its *Declaration on Euthanasia* we read: "By euthanasia is understood an action or omission of an action which of itself or by intention causes death in order that all suffering may be eliminated" (CDF, 1980a). And in *Evangelium Vitae* we read from John Paul

II that "Euthanasia is a violation of the law of God, since it is the deliberate and morally unacceptable killing of a person" (n. 65): "Thou shall not kill" (Fifth Commandment; cf. Ex. 20:13).

Today, too many terminally ill patients are being euthanized before they have come to a stage of acceptance and peace. Too many people are being put to death in times of anger, loneliness, and depression. A great injustice is being done to such people, all in the name of compassion.

The Church in its respect for the dignity of human life, and in its respect for God as the living Creator, promotes a holy death, a holy "letting go" which is filled with acceptance, peace, and hope on the part of the person entering into eternity.

The Church supports palliative care; that is, a form of care which seeks to eliminate pain and understands the redemptive value of unavoidable suffering (CCC 2279; cf. Col. 1:24). The Church therefore strongly encourages the use of painkillers in alleviating suffering, for at no stage is the "ordinary care owed to a sick person...[to be] interrupted" (CCC 2279). And for whatever pain remains, the Church encourages one to unite that suffering with Christ's for the good of one's soul and the souls of those in purgatory (cf. Col. 1:24).

When we euthanize people, we euthanize Jesus:

> All the nations will be gathered before him, and he will separate people one from another as a shepherd separates the sheep from the goats, and he will put the sheep at his right hand and the goats at the left. Then the king will say to those at his right hand, 'Come, you that are blessed by my Father, inherit the kingdom prepared for you from the foundation of the world; for I was hungry and you gave me food, I was thirsty and you gave me something to drink, I was a stranger and you welcomed me, I was naked and you gave me clothing, I was sick and you took care of me, I was in prison and you visited me.' Then the righteous will answer him, 'Lord, when was it that we saw you hungry and gave you food, or thirsty and gave you something to drink? And when was it that we saw you a stranger and welcomed you, or naked and gave you clothing? And when was it that we saw you sick or in prison and visited you?' And the king will answer them, 'Truly I tell you, just as you did it to one of the least of these who are members of my family, you did it to me.' Then he will say to those at his left hand, 'You that are accursed, depart from me into the eternal fire prepared for the devil and his angels; for I was hungry and you gave me no food, I was thirsty and you gave me nothing to drink, I was a stranger and you did not welcome me, naked and you did not give me

clothing, sick and in prison and you did not visit me.' Then they also will answer, 'Lord, when was it that we saw you hungry or thirsty or a stranger or naked or sick or in prison, and did not take care of you?' Then he will answer them, 'Truly I tell you, just as you did not do it to one of the least of these, you did not do it to me.' And these will go away into eternal punishment, but the righteous into eternal life' (Mt. 25: 32-46; NRSV).

When we euthanize people, we euthanize Jesus!

Euthanasia is a direct attack on God. It is an attack on his image and likeness (Gen. 1:27), his body (1 Cor. 12:12f; Rom. 12:5; Eph. 1:22f), his temple (1 Cor. 3:9-10, 15-16). Anyone who engages in euthanasia is killing the image and likeness of God, the Body of Christ, the Temple of God. They are committing an act of sacrilege, for we are "not our own" (cf. 1 Cor. 19-20)—our bodies belong to God!

Genetic engineering and assisted reproduction

Scientific research that aims at eliminating or overcoming sterility is of great merit as long as it seeks to maintain the unitive and procreative dimensions of the sexual act. It is gravely immoral to separate a husband from his wife (and vice versa) by introducing a third person into the reproductive process.

But that is exactly what secularism promotes in the name of self-interest, self-centeredness, self-pleasuring, and emotional comfort. In the name of secularism society is harmed because the very nature or essence of the family—which secularism implicitly denies—is disastrously manipulated and quite often destroyed. Children become possessions.

Immoral forms of genetic engineering and assisted reproduction are an affront to the image and likeness of what we were created to be like, the image and likeness of God (Gen. 1:27). It is an affront to the call to be Christ-like, to be the very Body of Christ (cf. 1 Cor. 12:12f; Rom. 1:22f), and it is an affront to God who dwells within us (1 Cor. 3:9-10, 15-16). It is a failure to recognize that we do not own our bodies but are stewards of our bodies—our bodies belong to God (cf. 1 Cor. 6:19-20).

Inappropriate forms of genetic engineering and assisted reproduction are an assault on God's creative, providential will and the dignity of marriage. In Matthew 19:5-6 we are reminded of God's will: "A man shall leave his father and mother and be joined to his wife, and the two shall become one flesh. They are no longer two, but one flesh. What God has joined together, no human being must separate." Just as Christ's union between his Body, the Church, cannot be separated (Eph. 5:22-32), likewise the union between husband and wife, a union which mirrors the relationship between Christ and his Church cannot be separated. Genetic engineering and assisted reproduction separate husband from wife and prevent two from becoming one. It thus blurs the nature of marriage, of love, and of Christ's Church. It blurs the very nature of God's relationship with his people (cf. Song of Songs).

It is an attack on the mystery of life—a diabolical attempt to control life as opposed to serve as its steward. The beauty of life becomes lost through man's inappropriate manipulation of God's work. When we read the following Scriptures we cannot help but shed a tear at what people are capable of doing in the name of improvement: In Genesis 25: 22-24 we read: "The children in Rebekah's womb jostled each other so much that she exclaimed, 'If this is to be so, what good will it do me!' She went to consult the Lord, and he answered her: 'Two nations are in your womb....'" In Jeremiah 1:5 we read: "Before I formed you in the womb I knew you, before you were born I dedicated you, a prophet to the nations I appointed you." In Isaiah we read: "Thus says the Lord who made you, who formed you from the womb: Fear not, O Jacob, my servant whom I have chosen" (v. 2 and v. 24). In Isaiah 49:2 we read: "The Lord called me from birth, from my mother's womb he gave me my name." In Job 10:8, 11 we read: "Your hands have formed me and fashioned me; with skin and flesh you clothed me, with bones and sinews you knit me together." And in Job 31:15 we read: "Did not he who made me in the womb make him? Did not the same One fashion us before our birth." In Psalm 139:13-16 we read: "You formed my inmost being; you knit me in my mother's womb. I praise you, so wonderfully you made me, wonderful are your works! My very self you knew; my bones were not hidden from you, when I was being made in secret, fashioned as in the depths of the earth. Your eyes foresaw my actions; in your book all are written down; my days were shaped, before one came to be." In Ecclesiastes 11:5 we read: "Just as you know not how the breath of life fashions the human frame in the mother's womb, so you know not the work of God which he is accomplishing in the universe." In Luke 1:41-44 we read, "When Elizabeth heard Mary's greeting, the infant leaped in her womb, and Elizabeth, filled with the Holy Spirit, cried out in a loud voice and said, 'Most blessed are you among women, and blessed is the fruit of your womb. And how does this happen to me that the mother of my Lord should come to me? For at the moment the sound of your greeting reached my ears, the infant in my womb leaped for joy." And in Luke 1:36 we read: "Behold, Elizabeth, has conceived a son in her old age, and this is the sixth month for her..."

The mystery and sanctity of life cannot but be diminished by the human person's inappropriate manipulations of the wonders of creation.

Artificial insemination
In the Catholic document *Donum Vitae* II, 1, 5, 4 we read:

Techniques that entail the dissociation of husband and wife, by the intrusion of a person other than the couple (donation of sperm, or ovum, surrogate uterus), are gravely immoral. These techniques (heterologous artificial insemination and fertilization) infringe the child's right to be born of a father and mother known to him and bound to each other by marriage. They betray the spouses' right to become a father and a mother only through each other.

Techniques involving only the married couple (homologous artificial insemination and fertilization) are perhaps less reprehensible, yet remain morally unacceptable. They dissociate the sexual act from the procreative act. The act which brings the child into existence is no longer an act by which two persons give themselves to one another, but one that 'entrusts the life and identity of the embryo into the power of doctors and biologists and establishes the domination of technology over the origin and destiny of the human person. Such a relationship of domination is in itself contrary to the dignity and equality that must be common to parents and children.' Under the moral aspect procreation is deprived of its proper perfection when it is not willed as the fruit of the conjugal act, that is to say, of the specific act of the spouses' union... Only respect for the link between the meanings of the conjugal act and respect for the unity of the human being make possible procreation in conformity with the dignity of the person.

At the heart of sexuality is the inseparable bond between the unitive and procreative dimensions of the conjugal act. This reality can be a tremendous cross upon a couple that so much desires the gift of a child. It must be remembered that children are not property owed to a couple. No one has a "right to a child." Only the child has rights, "the right to be the fruit of the specific act of the conjugal love of his parents," and "the right to be respected as a person from the moment of conception."

Those who are unable to have children by moral means should be encouraged to become generative by their works of charity and to seek the alternative of adoption, the giving of a loving home for parentless children, children hungering for the love of parents. The care of the orphan is a moral obligation and gift (cf. Exodus, Deuteronomy, Job, Psalms, Isaiah, Jeremiah, Titus etc...).

Designer Babies

When one is able to clone or to select what sex, hair or eye color, intellect, body structure, and so forth by genetic engineering and the manipulation and choice of embryos one is going down a dangerous path.

Huge distortions in the gene pool—which is essential for a healthy population—and huge distortions in the balance of the sexes in the population are bound to occur—cultures that prefer male children (often poor countries) will be overpopulated with males and cultures that favor female children will lead to an overpopulation of females. Designer babies will lead to distorted populations susceptible to grave illnesses, because of the diminished gene pool and the imbalance of the sexes (modern day China is a perfect example of this).

The striking, unique and unrepeatable qualities that make each of us special and distinctively beautiful are at stake when a culture seeks to play God. A culture that flirts with manipulating the origins of life is a culture flirting with extinction.

Cloning

> *"We are going to be one with God. We are going to have almost as much knowledge and almost as much power as God."*
> Richard Sheed, National Public Radio, 1998

In theory, human cloning is a way of producing a genetic replica of a person without sexual reproduction.

Cloning occurs when the nuclear material from a cell of an organism's body (a somatic cell) is transplanted into a female reproductive cell (an oocyte) whose nuclear material has been removed or inactivated in order to produce a new, genetically identical organism.

Those who favor cloning argue that one could theoretically harvest cells, blood, tissues, and much needed organs such as hearts, livers and kidneys for therapeutic use.

These harvested "products" would be considered ideal for they would be immunologically matched—that is, they would eliminate the need for life-long immunosuppressive therapy (Ahmann, NCBQ, 2001).

At another level, cloning would provide a means for sterile couples to reproduce.

At a glance cloning may appear appealing to some but in reality it is radically evil. As the ethicist Hans Jonas has written, [human cloning] is the most despotic...and the most slavish form of genetic manipulation" (*Tecnica, medicina edetica*, 1997).

The *Pontificia Academia Pro Vita* in its "Reflections on Cloning" points out that human cloning would radically damage the meaning, rationality, and complimentarity of human reproduction:

- The unitive, bonding aspect of human sexual reproduction would be lost in cloning. The precious gift of sexual intercourse as a physical and spiritual act between a man and a woman would become non-existent. A woman in theory could take the nuclear material from a somatic cell from her body and fuse it into her own ovum and produce a genetic reproduction of herself without any need of a husband.

- The naturally occurring balance between the male and female sex in society as well as the natural structure of the family would inevitably become distorted. As the document "Reflections on Cloning" explains: It is conceivable that "a woman could [end up being] the twin sister of her mother, lack a biological father and be the daughter of her grandfather."

- Human life would become viewed more as a "product," an object to be harvested, rather than as a gift of love. Cloning would suppress personal identity and subjectivity at the cost of biological qualities that could be appraised and selected. Women would be exploited for their ova and their wombs, being seen simply in terms of their "purely biological functions."

- Cloning could lead to a loss of genetic variation in society, thereby making society vulnerable to catastrophic illnesses and genetic defects. Naturally occurring mutations would not be sufficient to assure genetic variation.

- Cloning would lead to a wide array of psychological problems, whereby one would be troubled by questions such as: Who is my father? Who is my mother? Do I even have a father and mother? Who am I? What am I? Where do I come from?

- Cloning could lead to even greater trauma in the lives of parents who have lost a beloved child. The assumption from some heartbroken parents would be that if they could only clone their dead child, they would somehow have him or her back again. But this is not the case. A cloned individual would have a different guiding life principle and a different cultural and environmental upbringing. This child would not be what they desired or intended. Abuse of that cloned child could soon follow.

- One's "quality of life" would become a surrogate for one's search for meaning and salvation. A culture that is already self-centered and selfish would become even more so. It would become an even more "I, me, mine" culture.

- Human cloning could be the ultimate expression of narcissism and hedonism. One could envision a world that desires to clone only the so-called "beautiful" people. And who determines who are the beautiful people? Furthermore, one could envision a society in which a self-absorbed person would clone himself or herself so as to have spare parts in the event of illnesses.

- And finally, but most importantly, cloning would assault the dignity of human life in the most cruel and exploitative way imaginable by making cloned children the subject of experiments and by preventing their births.

Richard Sheed's words echo ominously: "We are going to become one with God. We are going to have almost as much knowledge and almost as much power as God." Cloning is an experiment in playing God. And we all know what happened in the story of Adam and Eve when they attempted to play God.

An Often Overlooked Reality of Cloning, Embryonic Stem Cell Research, and Invitro-Fertilization

One of the often overlooked evils associated with the above practices is that in the process of cloning, or doing embryonic stem cell research (see below), or attempting to have a child by means of artificial insemination, embryos are exploited and killed during the process—often in astronomical numbers. Human beings become biological debris. The commandment of "Thou shall not kill" is made into a farce!

Failure of Respect for Motherhood and Womanhood

Failure to respect the dignity of the human person from conception to natural death ultimately leads to the disintegration and death of a culture. When womanhood, motherhood, sexuality and procreation are distorted or lost, a culture cannot help but disintegrate, for the very foundation of culture, motherhood, is destroyed. What follows is a culture where divorce is viewed no more seriously than buying a new car. Promiscuity and the hatred for women inevitably follow because the nature of woman has been perverted. What follows is a culture where the bond of husband and wife, children and parents are lost.

Embryonic stem cell research

Stem cells are cells that have not undergone maturation and theoretically can become any of the 220 cell types and any of the 210 specialized tissue types that make up the human body.

Because stem cells are like "blank slates," they theoretically can morph into any kind of human tissue. They theoretically can become replacement parts for unhealthy cells and tissues. The benefits from stem cell research provides the future with great possibilities in the cure and treatment of illnesses, such as Parkinson's, Alzheimer's, heart disease, and diabetes.

Stem cells can be obtained immorally by the destruction of human life (i.e., human embryos) or they can be obtained morally from adults in a safe manner (i.e., from muscles, umbilical cords, bone marrow, the placenta, and from a wide variety of other adult tissues).

While there are no examples of success with embryonic stem cell research (except the killing of human life), great success has been attained in the use of adult stem cells. Adult stem cells not only have a future in curing and treating illnesses, they are doing so right now. Adult stem cells are currently being used in the treatment of multiple sclerosis, lupus, rheumatoid arthritis, stroke, anemia, Epstein-Barr virus infection, cornea damage, blood and liver diseases, brain tumors, retinoblastoma, ovarian cancer, solid tumors, testicular cancer, leukemia, breast cancer, neuroblastoma, non-Hodgkins' lymphoma, renal cell carcinoma, diabetes, heart damage; as well as cartilage, bone, muscle, and spinal-cord damage (NCCB, Life Issue Forum, 2001; *Science*, April, 2001; *Lancet*, January 2001; *APR*, 2000).

Given the benefits of adult stem cells, the question must be asked: Why are so many individuals preoccupied with embryonic stem cell research which involves the destruction of human life? Given the success of adult stem cells, you would think that these individuals would want improved funding and research in the field of adult stem cell experimentation.

The Scriptural objections and philosophical objections to embryonic stem cell research are obvious from the previous sections of this book.

XI
CONCLUDING REMARKS

There are no new questions!

The Catholic Church is more than 2000 years old. It has heard every question and has dealt with them all. There are no new questions. Therefore, as a Catholic, never feel fearful about your faith. If someone should ever come to you with what appears to be an absolutely perfect argument, do not get discouraged. It has been argued before, and it has been answered before. When such situations come up, all you need to do is to go to a good Catholic reference book and you will find the answer. In fact, you can even go to a good non-religious encyclopedia and find most of the answers you need to find.

Furthermore, let us never forget what Ignatius of Antioch (a disciple of the apostle John, and a bishop by the authority of Peter and Paul) said in his letter to the *Smyrnaeans*: "[Wherever] Jesus Christ is, there is the Catholic Church" (8).

And for those who seek to persecute the Catholic Church let them be reminded of what awaits them by the words of Lactantius (ca. 316) in his treatise on the *Deaths of the Persecutors*:

> *When Nero was already reigning Peter came to Rome, where, in virtue of the performance of certain miracles which he worked by the power of God which had been given him, he converted many to righteousness and established a firm and steadfast temple to God. When this fact was reported to Nero, he noticed that not only at Rome but everywhere great multitudes were daily abandoning the worship of idols, and, condemning their old ways, they were going over to the new religion. Being that Nero was a detestable and pernicious tyrant, he sprang to the task of tearing down the heavenly temple and of destroying righteousness. It was he that first persecuted the servants of God. Peter, he fixed to a cross upside down; and Paul he beheaded (2, 5). [For his persecution of the Church, Nero would pay with his life].*

Those who persecute the Church will always fail as they have always failed!

The future of Christianity

What does the future hold? Obviously only God knows. But we can certainly speculate.

The Eastern Orthodox Churches will continue to come back home to Rome, to the successor of Peter. The Orthodox Churches already recognize him as having primacy of honor among all the bishops, since he is the successor of St. Peter, the head of the apostles. The only stumbling block remaining is the affirmation of Rome in terms of primacy of jurisdiction. This is a minor issue which will be solved in our lifetime. As mentioned above, the Orthodox Churches are coming home to Rome in record numbers. Once the Greek and Russian Orthodox Churches reunite with Rome, the Church will have all the successors of the apostles under one roof.

What about the 33,000 Protestant denominations and 150,000 pseudo-Protestant denominations? Since Protestants lack what is necessary to make a Church, apostolic succession, they are referred to as ecclesiastical communities. And as ecclesiastical communities they will continue to split (150,000 times since the 16th century alone), since they are not firmly fastened to that pillar of truth which comes with apostolic succession, that protection that comes from a line of successors, of bishops, that trace themselves to Jesus Christ himself. Thus Protestantism will continue to split and disintegrate into nothingness. It will eventually become more secular than religious. It will become more worldly than Godly. Protestantism will become more of a collection of social clubs than religious denominations. All Protestantism will become non-denominational in practice if not in name.

As Christians become more educated in the history of Christianity and doctrine, of the history of the formation of the Bible, they will inevitably return to Rome.

Protestantism will continue to become a faith of personal opinion and personal belief. It will continue to progress towards being a faith where "no absolutes" will be recognized—either at an implicit level or an explicit level. All Protestantism will become non-denominational in practice if not in name.

With time, all Christians of substance will return to the successor of Peter, the Pope.

This will mark the beginning of the battle of all battles—the battle of believers against secularists, relativists, and pagan worshipers of the culture of death. The superficial Christian will disappear. The Christian of the future will end up being a mystic in the world or nothing at all.

This purified Church will win the day as Christ returns in glory.

The ship continues moving!

The Church is like a ship moving towards heaven. One can get on board, stay on board, or get off. But the ship will keep moving forward, with you or without you, with me, or without me. The Church has been sailing for more than 2000 years. Some have abandoned the ship and some have embarked. The ship will always go on. The gates of hell will never prevail against it (Mt. 16:18; 28:20). Are you on board?

APPENDIX I
AN APOLOGETICS DEBATE HANDBOOK

Word of God
Bible, Tradition and Magisterium
*(Three inseparable realities necessary for knowing the Word of God—
for properly interpreting the Word of God)*

Sacred Scripture

Jer. 30:1-3: inspired by God....
2 Tim. 3: 16-17: useful for teaching, correction...

Sacred Tradition

2 Jn. 1:12: do not intend...paper and ink
Jn. 16:12-13: I have much more to tell you, but you cannot bear it now
1 Cor. 11:2: hold fast to traditions I handed on to you
1 Cor. 15:1-3: being saved if you hold fast to the word I preached
2 Thess. 2:15: hold fast to traditions, whether oral or by letter
2 Thess. 3:6: shun those not acting according to tradition
Jn. 21:25: whole world could not contain the books of Jesus' words
2 Tim. 1:13: take as your norm the sound words that you heard from me
2 Tim. 2:2: what you heard from me entrust and teach to others
1 Pet. 1:25: God's eternal word=preached to you
Rom. 10:17: faith comes from what is heard
Mk. 16:15: go to the whole world and proclaim gospel

Origen (230 AD): "The teaching of the Church has indeed been handed down through an order of succession from the apostles, and remains in the churches even to the present time. That alone is to be believed as the truth which is in no way at variance with ecclesiastical and apostolic tradition." *Fundamental Doctrines 1, preface 2.*

Sacred Tradition vs. Human Traditions
Human Traditions are Condemned

Sacred Tradition is that which comes from the apostles and was carried on by the successors of the apostles in faith and morals. Human tradition is not based on faith and morals, but on human whims detached from the apostles and their successors.

Mt. 15:3: break commandments of God for your human traditions

Mk. 7:9: set aside God's commandments for human tradition
Col. 2:8: seductive philosophy according to human tradition

Magisterium
Magisterium—apostles and their successors, the bishops

2 Pet. 1:20: no prophecy is a matter of private interpretation
2 Pet. 3:15-16: Paul's letters can be difficult to grasp and interpret
Acts 15:1-21: apostles meet in a council to interpret and set the doctrine
of belief for the Gentiles
Acts 8:26-40: the apostle Philip is needed to authentically interpret the
Scriptures to the Ethiopian eunuch

Scripture, Tradition, and the Magisterium
Vincent of Lerins (ca. 450) in his *Commonitoria* (2,1-3) beautifully
illustrates the need for Sacred Tradition, Sacred Scripture, and the teaching
office of the Church (the Magisterium) when seeking the authentic word of
God.

*With great zeal and closest attention...I frequently inquired of many
men eminent for their holiness and doctrine, how I might, in a
concise and, so to speak, general and ordinary way, distinguish the
truth of the Catholic faith from the falsehood of heretical depravity.
I received almost always the same answer from all of them, that if I
or anyone else wanted to expose the frauds and escape the snares of
the heretics who rise up, and to remain intact and sound in a sound
faith, it would be necessary, with the help of the Lord to fortify that
faith in a [pertinent] manner: first, of course, by the authority of the
divine law; and then, by the Tradition of the Catholic Church. Here,
perhaps, someone may ask: 'If the canon of the Scriptures be
perfect, and in itself more than suffices for everything, why is it
necessary that the authority of ecclesiastical interpretation be joined
to it?' Because, quite plainly, Sacred Scripture, by reason of its own
depth, is not accepted by everyone as having one and the same
meaning. The same passage is interpreted by others, so that it can
almost appear as if there are as many opinions as there are men.
Novatian explains a passage in one way, Sabellius another, Donatus
in another; Arius, Eunomius, Macedonius in another; Photinus,
Apollinaris, Priscillian in another; Jovinian, Pelagius, Caelestius in
another.... [Without reference to the Tradition as expounded and
taught by the apostles and their successors, the bishops, there would
be no way of knowing the true meaning of the Scriptures.]*
(Jurgens, vol. 3).

The Bible "only" approach to divine revelation is unbiblical and contrary to Sacred Tradition—which we are commanded to hold onto (2 Thess. 2:14-15). The Bible "only" approach is a "human" tradition or invention which is contrary to the deposit of the faith (Matt. 15:3, 6-9; Col. 2:8).

Nature of the Church

1 Cor. 3:11: Church founded by Christ
1 Cor. 6:15; 12:12-27; Rom. 12:5; Eph. 1:22f; 5:30: body of Christ
1 Cor. 3:9-10, 16: God's building and Temple
2 Cor. 11:2; Eph. 5:25, 27, 29; Rev. 19:7: bride of Christ
Lk. 12:32; Jn. 10:3-5, 11: flock of Christ
Mt. 16:18f; 28:19-20: Church will last forever
Eph. 4:11-16: possesses the means of salvation

One true Church

Jn. 10:16: there shall be one fold and one shepherd
Jn. 17:17-23: I pray that they may be one, as we are one
Eph. 4:3-6: there is one Lord, one faith, one baptism, one God and Father
1 Cor. 12:13: in one spirit we were baptized into one body
Col. 3:15: the peace into which you were called in one body

Tradition

Cyprian of Carthage (250 AD): "God is one and Christ is one, and one is his Church, and the faith is one…. Unity cannot be rent asunder, nor can the one body of the Church be divided into separate pieces" *On the Unity of the Church.*

Infallible in faith and morals

Mt. 16:18-19: upon this rock (kepa) I will build my Church
Mt. 18:17-18: if he refuses to listen even to the church/ power to legislate and discipline
Mt. 28:18-20; Jn. 20:23; 1 Cor. 11:24: power delegated to apostles and successors/ I am with you always
Lk. 10:16: whoever hears you, hears me; rejects you, rejects me/ speaking with Christ's voice and authority
Jn. 14:26: Holy Spirit to teach and remind them of everything
Jn. 16:12-13: Spirit of truth will guide you to all truth
1 Tim. 3:15: Church is the pillar and foundation of truth

<u>Acts 15:28</u>: apostles speak with voice of Holy Spirit

Tradition

Irenaeus (200 AD): "For where the Church is, there is the Spirit of God; and where the Spirit of God, there the Church and every grace." *Against Heresies 3, 24, 1.*

Infallibility tied to apostolic succession

<u>Jn. 20:21</u>: Jesus gave the apostles his own mission
<u>1 Tim. 3:1, 8; 5:17</u>: identifies roles of bishops, presbyters
  (priests) and deacons to govern his Church
<u>1 Tim. 4:14</u>: gift of ordination conferred by the laying on of the hands
<u>1 Tim. 5:22</u>: do not lay hands for ordination too readily
<u>Acts 1:15-26</u>: Matthias is chosen to succeed Judas—apostolic succession
<u>Acts 14:23</u>: they appointed presbyters in each community
<u>Tit. 1:5</u>: commission for bishops to ordain priests—succession of
  authority

Tradition

In seeking where to find the true faith we must say to those who claim to have it the following: "Unroll the order of your bishops, running down in succession from the beginning, so that your first bishop shall have for author and predecessor one of the apostles."

> *Tertullian,*
> *The Demurrer Against Heresies* (ca. 200) (32:1)

There is no true Church without apostolic succession!

Infallibility tied to primacy of Peter

<u>Mt. 16:18-19f</u>: upon this rock/Peter/kepa I will build my Church with
  power to bind and loose
<u>Lk. 9:32; Mk. 16:7</u>: "Peter and his companions"
<u>Mt. 18:21; Mk. 8:29; Lk. 8:46; 12:41; Jn. 6:68-69</u>: spoke for apostles
<u>Lk. 22:32</u>: Peter's faith will strengthen his brethren
<u>Jn. 21:17</u>: given Christ's flock as chief shepherd
<u>Acts 1:13-26</u>: headed meeting which elected Matthias
<u>Acts 2:14</u>: led apostles in preaching on Pentecost
<u>Acts 2:41</u>: received first converts

Acts 3:6-7: performed first miracle after Pentecost
Acts 5:1-11: inflicted first punishment: Ananias and Saphira
Acts 8:21: excommunicated first heretic, Simon Magnus
Acts 10:44-46: received revelation to admit Gentiles into Church
Acts 15: led first Church counsel
Acts 15:7-11: pronounces first dogmatic decision
Gal. 1:18: after conversion, Paul visits Peter
Mt. 10:1-4; Mk. 3:16-19; Lk. 6:14-16; Acts 1:13: Peter's name always heads list of apostles

Tradition

In this chair in which he himself sat, Peter,
In mighty Rome, commanded Linus, the first elected, to sit down.
After him, Cletus too accepted the flock of the fold.
As his successor, Anacletus was elected by lot.
Clement follows him, well-known to apostolic men.
After him Evaristus ruled the flock without crime.
Alexander, sixth in succession, commends the fold to Sixtus.
After his illustrious times were completed, he passed it on to Telesphorus
He was excellent, a faithful martyr.
After him, learned in the law and a sure teacher,
Hyginus, in the ninth place, now accepted the chair.
Then Pius, after him, whose blood-brother was Hermas,
An angelic shepherd, because he spoke the words delivered to him;
And Anicetus accepted his lot in pious succession.

Tertullian (ca. 193)
Adversus Marcionem libri quinque , 3, 276-285; 293-296

And in Cyprian of Carthage (ca. 251) we read in *De Ecclesiae Unitate* (cf. 2-7):

The blessed apostle Paul teaches us that the Church is one, for it has 'one body, one spirit, one hope, one faith, one baptism, and one God.' Furthermore, it is on Peter that Jesus built his Church, and to him he gives the command to feed the sheep; and although he assigns like power to all the apostles, yet he founded a single chair, and he established by his own authority a source and an intrinsic reason for that unity. Indeed, the others were that also which Peter was; but a primacy is given to Peter, whereby it is made clear that there is but one Church and one Chair—the Chair of Peter. So too, all are shepherds, and the flock is shown to be one, fed by all the apostles in single-minded accord. If someone does not hold fast to this unity of Peter, can he imagine that he still holds the faith? If he deserts the chair of Peter upon whom the Church was built, can he still be confident that he is in the Church?

In his *Letter to all his People* [43 (40) 5] written in 251 AD Cyprian reminded his people that the faith of the pope was the faith of the Church:

They who have not peace themselves now offer peace to others. They who have withdrawn from the Church promise to lead back and to recall the lapsed to the Church. There is one God and one Christ, and one Church, and one Chair founded on Peter by the word of the Lord. It is not possible to set up another altar or for there to be another priesthood besides that one altar and that one priesthood. Whoever has gathered elsewhere is scattering.

The 264th successor of Peter, the 265th pope is Benedict XVI, now reigning.

The Trinity, Father, Son, and Holy Spirit

Trinity

Gen. 1:26: "Let *us* make man in *our* image, after *our* likeness"
Gen. 3:22: "the man has become like one of *us*"

Elohim—plural noun for God, a oneness and a plurality

Mt. 3:16f: voice, Jesus, image of dove—at Jesus' baptism
Mt. 17:5: Transfiguration—voice, beloved one, cloud/shadow
Mt. 28:19: baptize in the name of the Father, the Son, ...
2 Cor. 13:13: "grace of the Lord Jesus Christ and the love of God and the fellowship of the Holy Spirit be with you"

OT—Father NT—Son ACTS—Holy Spirit
211 AD—Tertullian/ sign of the cross

Gen./ Heb. 5:5: Father eternally generates
Jn. 1:1-4f: Son is eternally begotten
Jn. 15:26: Holy Spirit eternally proceeds

Analogy: water as liquid, gas, solid/ sun, ray, light

Christ

Mt. 1:23: "Emmanuel...God is with us."

Ex. 3:14; Jn. 8:58; Jn. 18:4-8: "I AM": 11 times in John's Gospel alone

Col. 2:9: "In Christ the fullness of deity resides in bodily form"

Jn. 10:30; 14:9-11; 17:11; 17:21: the Father and I are one

Jn. 20:28: "my Lord (Yahweh) and my God (Elohim)"

Jn. 20:16: Rabbuni—used usually to address God

2 Tim. 4:18; 2 Pet. 3:18; Rev. 1:6; Heb. 13:20-21: phrase "to him be glory for ever" usually reserved for God

Mt. 5:1-12: Sermon on the Mount. Jesus is the New Moses, but greater. Whereas Moses received the commandments and then brought them to the people, Jesus gives the law to the people from the mount directly. Jesus is God as the lawgiver from the mount as God was the lawgiver to Moses from Sinai. Jesus is God and the New Moses!

Mt. 19:28; 25:31; Jn. 5:22; Acts 10:42: judging living and dead, a divine prerogative

Mt. 9:6: forgiveness of sin, a divine prerogative

Jn. 1:3; Col. 1:16f; Heb. 1:2: creator of all things, a divine prerogative

Mk. 1:1; Lk. 1:32; Jn. 1:34: Son of God

1 Cor. 2:8: Lord of Glory

Rev. 17:14; 19:16: King of Kings

Rev. 1:8f: Alpha and Omega

Mt. 1:21; Jn. 3:14-15: Savior

Jn. 1:1-5, 14; Mt. 1:23; Lk. 2:52: assumed a human nature

Jn. 4:34; Mk. 14:36: divine and human will

Jn. 14:28: "the Father is greater than I"—in Jesus' human nature the Father is greater; in Jesus' divine nature, he is equal to the Father. Why is this? See next quote from Philippians 2:7f. Emptied himself of all except what was necessary for our salvation.

Phil. 2:7f: assumed a human, rational soul

1 Pet. 2:22; Jn. 8:46; 2 Cor. 5:21; Heb. 4:15: immune from sin

Holy Spirit

Acts 5:3-4: lie to Holy Spirit is a lie to God

Jn. 15:26: proceeds from the Father

Lk. 1:35: makes us aware of the Incarnation

Jn. 20:22-23: the forgiveness of sins

1 Cor. 6:11 Rom. 15:16: justification and sanctification

Rom. 5:5: charity of God

Jn. 14:16-17; 15:26: spirit of truth

Acts 6:5: strengthens our faith

Rom. 8:9-11; 1 Cor. 3:16; 6:19: dwells within us

Acts 8:29: guides our works
1 Cor. 12:11; 1 Cor. 12:4-11: supernatural gifts and supernatural life
Cf. Jn. 14:16-18; Acts 5:3f; 1 Cor. 2:10f; 3:16; 6:11, 19f; 1 Pet. 1:1-3;
   Eph. 4:4-6: attests to his divinity and consubstantiality or oneness
   with the Father and the Son

Is. 11:1-2: gifts of spirit—wisdom, understanding, counsel, fortitude,
   knowledge, piety, fear of the Lord
Gal. 5:22-23: fruits of the spirit—love, joy, peace, patience, kindness,
   generosity, faithfulness, gentleness, and self-control

---

*When put together they point to Jesus as fully human, fully divine,
without confusion, division, nor separation between the two natures*

Tradition

The Fathers and Early Writers and the Trinity

[In the early Church, according to Pliny, the Roman Governor of Pontus, in
his Letters to the Emperor Trajan (ca. 111-113 AD,) the Christian faithful
would often sing a "hymn to Christ as God" as they began their celebration
of the "Lord's Supper."]

> *Christians are brought to future life by one thing...that they recognize
> that there is a oneness, a unity, a communion between the Son and the
> Father, and that there is a oneness, a unity, a communion, albeit a
> distinction, between the Spirit, the Son, and the Father.*
> *Justin Martyr (ca. 148 AD), Legat. Pro Christ*

Seven Sacraments
Baptism

Original sin
   Gen. 2:16-17: the day you eat of that tree, you shall die
   Gen. 3:11-19: God's punishment for eating of the tree
   Rom. 5:12-19: many became sinners through one man's sin
   1 Cor. 15:21-23: by a man came death; in Adam all die
   Eph. 2:1-3: we all once lived in the passions of our flesh

Baptism
   Mt. 28:19; Mk. 16:16; Jn. 3:5: commanded by Christ

Acts 2:38, 41; 8:12, 38; 9:18; 10:48: taught and administered by the apostles

Mk. 1:4, 8; Jn. 1:33; 3:5; Tit. 3:5: laver of regeneration

Mk. 16:16; Acts 2:38; 8:12f; 16:33; Rom. 6:3-6; Gal. 3:27; 1 Cor. 6:11; Eph. 5:26; Col. 2:12-14; Heb. 10:22: takes away all sin

Acts 2:38; 19:5f: receive the Holy Spirit

Gal. 3:27; Mk. 10:38; Lk. 12:50; Rom. 6:3: put on Christ, and are baptized into his death and resurrection

1 Cor. 12:13, 27: enter into the Church

Of Infants

Rom. 5:18-19: all are born of Adam's sin and need baptism

Jn. 3:5/ Mk. 16:16: baptism necessary for salvation

Mk. 10:14: let the children come; to such belongs the kingdom

Lk. 18:15: people were bringing even infants to him

Acts 16:15: she was baptized, with all her household

Acts 16:33: he and all his family were baptized at once

1 Cor. 1:16: I (Paul) baptized the household of Stephanas

Col. 2:11-12: baptism replaces circumcision (which was done on the eighth day) for being part of the people of God, of the new covenant. Old Testament infants did not make a choice regarding their circumcision, their being part of the people of God; their parents did—likewise in baptism

Mt. 8:5ff/ Mt. 15:21f

Faith of parents speaks for infants/children (i.e., Servant was healed through centurion's faith/ daughter was healed through mother's faith

Tradition

Hippolytus of Rome, (215 AD), *Apostolic Tradition*, 21: "Baptize first the children. Let their parents or other relatives speak for them."

Origen (244 AD), *Commentary on Romans*: "the Church received from the apostles the tradition of giving baptism also to infants."

Confirmation
Completes baptism

Acts 19:5-6: Paul imposed hands on baptized & they received Holy Spirit

Acts 8:14-17: laid hands upon them; they received the Holy Spirit

2 Cor. 1:21-22: put seal on us/ given Spirit in our hearts

Eph. 1:13: you were sealed with the promised Holy Spirit

Heb. 6:2: instruction about baptism and laying on of hands

Tradition

Cyril of Jerusalem (350 AD) beautifully summarizes the power and the necessity of the Sacrament of Confirmation:

> *And to you in like manner, after you had come up from the pool of the sacred streams, there was given chrism, and this is the Holy Spirit (21 [3] 1). But beware of supposing that this is ordinary ointment. For just as the bread of the Eucharist after the invocation of the Holy Spirit is no longer simple bread, but the Body of Christ, so also this holy ointment is no longer plain ointment, nor, so to speak, common, after the invocation. Rather, it is the gracious gift of Christ; and it is made fit for the imparting of his godhead by the coming of the Holy Spirit. This ointment is applied to your forehead and to your other senses; and while your body is anointed with the visible ointment, your soul is sanctified by the Holy and life-creating Spirit (21 [3] 3). Just as Christ, after his baptism and the coming upon him of the Holy Spirit went forth and defeated the adversary, so also with you; after holy baptism and the mystical chrism of the [Sacrament of Confirmation] and the putting on of the panoply of the Holy Spirit, you are able to withstand the power of the adversary and defeat him by saying, 'I am able to do all things in Christ who strengthens me (21 [3] 4) (Mystagogic).*

Holy Eucharist

Mal. 1:11: prophesied—"Gentiles" offering sacrifice!!

Ex. 16:15: prefigured/ *man hu* which has its origins in the Hebrew word *"man"* but often rendered *"manna."* The "unknown" food from heaven!

Mt. 26:26-28; Mk. 14:22-24; Lk. 22:19f: "Do this in remembrance of me"

1 Cor. 10:16: Eucharist—participation in Christ's body and blood

1 Cor. 11:23-29: receiving unworthily=guilty of his body and blood

~Ex. 12:8, 46: Paschal Lamb had to be eaten

~Jn. 1:29: Jesus is Lamb of God—therefore to be eaten

~1 Cor. 5:7: Jesus is "paschal lamb who has been sacrificed"

Jn. 6:35-71: Real Presence

Gen. 9:3-4; Lev. 17:14: eating food and drinking blood forbidden/ explains why they abandoned Jesus

Jn. 6:63: requires gift of God to understand (cf. Jn. 3:6)

- There are several dimensions to the Eucharist, one is symbolic (to do the will of Father/ the gift of faith), the other is "real presence" and "real sacrifice" (Jn. 4:31-34; Mt. 16:5-12)

- Ignatius of Antioch (110 AD), _Letter to the Smyrneans_, 6, 2: "heretics abstain from the Eucharist because they do not confess that the Eucharist is the **Flesh** of our Savior Jesus Christ."
- Justin Martyr (150 AD), _First Apology_, 66,20: Not as common bread nor common drink do we receive these; but as we have been taught, the food which has been made **into** the Eucharist by the Eucharistic prayer set down by Christ, and by the change of which our blood and flesh is nourished, is both the **Flesh and Blood of that Incarnated Jesus**."
- Irenaeus of Lyons (195 AD), _Against Heresies_, 5,2,2: "Jesus has declared the cup, a part of his creation, to be **His own Blood**, from which he causes our blood to flow, and the bread, a part of creation, he has established as **His own Body**, from which he gives increase to our bodies.
- Cyril of Jerusalem (350 AD), _Catechetical Lectures, Mystagogic_ 4,22,1, 6: "He himself having declared and said of the bread, "**This is My Body**," who will dare any longer to doubt? And when he himself has affirmed and said, "**This is My Blood**," who can ever hesitate and say it is not His Blood?".... "Do not therefore regard the bread and wine as simply that, for they are, according to the Master's declaration, the Body and Blood of Christ. Even though the senses suggest to you the other, let faith make you firm, not doubting that you have been deemed worthy of the Body and Blood of Christ."

Penance

James 5:16: confess your sins to one another—_not simply to God_
James 5:13-15: prayer of presbyters forgives sins
Mt. 18:18; 16:19: whatever you bind and loose on earth...in heaven
Jn. 20:22-23: "Jesus breathed on them and said to them, 'Receive the Holy Spirit.' Whose sins you forgive are forgiven them, and whose sins you retain are retained."
"Receive the Holy Spirit" is a reference to Genesis 2:7 where God breathed life into man. In the forgiveness of sins, we have new life!

Mortal and Venial Sins
1 Jn. 5:16-17: some sins are deadly, some sins are not deadly

Tradition

Theodore of Mopsuestia (ca. 428) reminds us that the priest is a father (as in 1 Cor. 4:14-15; 1 Tim. 1:2; Tit. 1:4; Philem. 10; 1 Thess. 2:1) who takes care of his children, a spiritual doctor that brings healing to souls:

> *If we commit a great sin against the commandments we must first induce our conscience with all our power to make haste and repent of our sins as is proper, and not permit ourselves any other medicine. This is the medicine for sins, established by God and delivered to the priests of the Church, who make diligent use of it in healing the affliction of men. You are aware of these things, as also of the fact that God, because he greatly cares for us, gave us penitence and showed us the medicine of repentance; and he established some men, those who are priests, as physicians of sins. If in this world we receive through them healing and forgiveness of sins, we shall be delivered from the judgment that is to come. It behooves us, therefore, to draw near to the priests in great confidence and to reveal to them our sins; and those priests, with all diligence, solicitude, and love, and in accord with the regulations mentioned above, will grant healing to sinners. The priests will not disclose the things that ought not to be disclosed; rather, they will be silent about the things that have happened, as befits true and loving fathers who are bound to guard the shame of their children while striving to heal their bodies... (Catechetical Homilies, 16).*

The Church has always, from the beginning of the Church, had confession of sins to priests.

Anointing of the Sick

James 5:14f: presbyters pray over sick, anoint and bring forgiveness of sins

Mk. 6:12-13: anointed the sick, many cured

Tradition

Origen writing in 244 AD affirms this biblical teaching on the Anointing of the Sick when he wrote:

> *Let the priests impose their hands on the sick and anoint them with oil and the sacrament will heal the sick persons and forgive them their sins.*
>
> *Homilies on Leviticus, 2, 7, 8*

Holy Orders

Lk. 22:19; Jn. 20:22f: *Instituted by Christ*
"Do this in memory of me." "As the Father sent me, I send you...receive the Holy Spirit."
Acts 6:1: ministry of an ordained deacon
1 Tim. 3:1-16: bishops and deacons/ qualifications
Rom: 16:1: Deaconess: name given historically to the wife of a deacon. (found only in RSV, NJB). "Deaconoi" also commonly used for ministers that are not ordained as well.
Acts 20:17-28: Presbyters summoned and told them that they have been appointed by the Holy Spirit as overseers of the Church
Acts 13:3: they laid hands on them and sent them off
Acts 14:23: they appointed presbyters in each church
Tit. 1:5: appoint presbyters in every town, as I directed you
give grace/ ordained by those ordained
1 Tim. 4:14: gift received through the laying on of hands by presbyters
2 Tim. 1:6: gift of God you have through imposition of hands

Father Quote

Mt. 23:9: Call no man your father?
Acts 7:2: Stephen calls Jewish leaders "fathers"
Acts 22:1: Paul calls Jerusalem Jews "fathers"
Rom. 4:16-17: Abraham called "the father of us all"
1 Cor. 4:14-15: I became your father in Christ
1 Tim. 1:2: my true child in the faith
Tit. 1:4: my true child in our common faith
Heb. 12:7-9: we have earthly fathers to discipline us
Philemon 1:10: whose father I became in my imprisonment
1 Jn. 2:13,14: I write to you fathers...

Celibacy

Mt. 19:12: celibacy praised by Jesus who was celibate
1 Cor. 7:8; Saint Paul was celibate
1 Cor. 7:32-35: celibacy is recommended for full-time ministers

1 Tim. 5:9-12: pledge of celibacy taken by older widow
Order of Religious and virgins

Tradition
Let the bishop be ordained after he has been chosen. When someone

pleasing to all has been named, let the people assemble on the Lord's Day with the presbyters and with such bishops as may be present. All giving assent, the bishops shall impose hands on him, and the presbytery shall stand in silence (2). When the presbyter is to be ordained, the bishop shall impose his hand upon his head while the presbyters touch the one to be ordained....(8). When a deacon is to be ordained the bishop alone shall lay his hands upon him (9).

Hippolytus of Rome (ca. 200)

Matrimony

How can I ever express the happiness of a marriage joined by the Church, strengthened by an offering, sealed by a blessing, announced by angels, and ratified by the Father? How wonderful the bond between two believers, now one in hope, one in desire, one in discipline, one in the same service! They are both children of one Father and servants of the same Master, undivided in spirit and flesh, truly two in one flesh. Where the flesh is one, one also is the spirit.

Tertullian (ca. 155-240)
Ad uxorem, 2, 8, 6-7: PL 1, 1412-1413

Mt. 19:5: leave father and mother, join wife, and two shall become one flesh
Mk. 10:7-12: what God has joined together, no man must separate
Eph. 5:22-32: union of man and wife as image of the inseparable nature between Christ and his Church
1 Thess. 4:4: acquire a wife for yourself in holiness and honor

Divorce and remarriage (without an annulment), contrary to God's will

Mal. 2:14-16: for I hate divorce, says the Lord
Mk. 10:11-12: if either divorces and remarries=adultery
1 Cor. 7:10-11: if wife separates, stay single or reconcile
Rom. 7:2-3: death frees one to remarry
as does annulments—see below

Annulments

An annulment is the recognition that a marriage was never blessed by God—that is, it was never elevated to the level of a sacrament. It was an "unlawful marriage."

Mt. 5:32-33; Acts 15:20; 15:29; Mt. 19:5-9: reference to "unlawful marriages"

Lev. 18: examples of some "unlawful marriages"

Mt. 14:3-12: King Herod's marriage to Herodias, Herod's brother Philip's wife, viewed as "unlawful"

Pauline Privilege: 1 Cor. 7:12-15

Marriage of two non-baptized persons is dissolved, when after the separation and divorce, one of the persons converts to Christianity.

Petrine Privilege: Marriage between a non-baptized person and a baptized person. Since it is not a sacramental marriage, it may be dissolved by virtue of Peter's right to bind and loose.

Contraception, contrary to God's Will for Marriage

Gen. 1:27-28: be fruitful and multiply

Gen. 38:9-10: Onan killed for spilling "his seed" on the ground

Rev. 9:21: the Greek word here for "magic potions" is *pharmakeia*—an abortion causing agent

Divorce and Remarriage, contrary to God's will

Divorce and remarriage (without annulment), adultery: see above section

Gay marriages contrary to God's will for marriage

Gen. 1: complimentarity of sexes as a reflection of God's inner unity

Gen. 2: transmission of life and love through self-donation and the becoming of one flesh—which implies a complimentarity between the sexes

Gen. 19: Sodom destroyed—for homosexual acts

Lev. 18:22, 29: practicing homosexuals are an abomination and must be cut off from the people of God

Lev. 20:13: practicing homosexuals were once put to death

Rom. 1:27: called unnatural, shameful and a perversity

1 Cor. 6:9: practicing homosexuals will be condemned to hell

1 Tim. 1:10: practicing homosexuals are sinners

Abortion, contrary to God's will for marriage

Gen. 25:22-24: two nations are in your womb

Jer. 1:5: before you were born I dedicated you.
Isaiah 44:2, 24: who formed you from the womb
Isaiah 49:2: from my mother's womb he gave me my name
Job. 10:8, 11: with bones and sinews you knit me together
Job. 31:15: fashioned us before our birth
Psalm 139:13-16: my days were shaped before one came to be
Ecclesiastes 11:5: breath of life fashions the human frame
Lk. 1:41-44: the infant leaped in her womb
Lk. 1:36: sixth month for her
Rev. 9:21: the Greek word here for "magic potions" is *pharmakeia*—an abortion causing agent

Tradition

"You shall not kill an unborn child or murder a newborn infant."
Didache, II, 2 (ca. 65 AD)

"You shall love your neighbor more than your own life. You shall not slay the child by abortion."
Barnabas, *Epistles*, II (ca. 70-138).

"For us murder is once and for all forbidden; so even the child in the womb, while yet the mother's blood is still being drawn on to form the human being, it is not lawful for us to destroy. To forbid birth is only quicker murder. He is a man, who is to be a man; the fruit is always present in the seed."
Tertullian, 197, *Apologetics* (ca. 197)

"Those who use drugs to bring about an abortion commit murder and will have to give an account to God for their abortion."
Athenagoras, *Legatio pro Christianis*, (ca. 177)

"There are women, who, by the use of medicinal potions, destroy the unborn life in their wombs, and murder the child before they bring it forth. These practices undoubtedly are derived from a custom established by your gods; Saturn, though he did not expose his sons, certainly devoured them."
Minucius Felix, *Octavius* (ca. 200)

"If we would not kill off the human race born and developing according to God's plan, then our whole lives would be lived according to nature. Women who make use of some sort of deadly abortion drug kill not only the embryo but, together with it, all human kindness."
Clement of Alexandria, *Christ the Educator, II* (ca. 150-220)

Mary

Mother of God listen to my petitions; do not disregard us in adversity, but receive us from danger.
                                                    Second Century Papyrus, Or. 24, II.

Mary's unique dignity
Lk. 1:42: "most blessed among women"
Lk. 1:48: "all generations will call me blessed"
Lk. 1:28, 30: "full of grace," found favor with God

Mother of God
Mt. 1:21-23: Emmanuel which means 'God is with us'
Lk. 1:43: Mother of my Lord

Mary without sin—Immaculate
Lk. 1:28, 30: "full of grace" the Lord is with you (*kecharitomene*) / Jesus is "full of grace" (Jn. 1:14).
Lk. 1:35: "to overshadow" Mary "to overshadow" the Ark of the Covenant (Ex. 40: 34-35)
Ex. 40: ark made perfect in every detail to allow God's presence to fill it

Tradition

In the year 306 AD we read in Ephraeim's *Nisbene Hymn* (27, 8) the following:

> *You alone and your Mother*
> *Are more beautiful than any others;*
> *For there is no blemish in you,*
> *Nor any stains or sins upon your Mother*

Without sin because of the Savior
Lk. 1:47: "my spirit rejoices in God my Savior"

What about Rom. 3:23? "All have sinned and are deprived of the glory of God" This is a generalization for the mass of humanity since Jesus never sinned and infants do not sin until they reach the age of childhood, the age of reason.

In Genesis 1:2f we are reminded that from the immaculately created cosmos God created Adam. In Romans 5:14 and 1 Corinthians 15:22 we are reminded that Jesus is the second Adam. If the first Adam was created from pristine organic materials, what would the second Adam be created from? Obviously an immaculate, pristine Mother!

Perpetual Virginity
Zech. 12:10: "as one mourns an only son"
Mary's children never mentioned

"brothers and sisters?"
What about Mt. 12:46, John 7:5; Acts 1:14; Gal. 1:19; Mark. 3:31-35; 6:3? In the Hebrew and Aramaic there was no word for cousin, uncle, nephew, niece, half-brother or sister, or close relative. The NT alone uses the word brother for all kinds of relationships 325 times. In Acts 1:15f Peter addresses the 120 brothers! James and Joseph are referred to as "the brothers of Jesus" and yet are "sons of another [the other] Mary, a disciple of Christ" (cf. Mt. 13:55; 28:1; cf. Mt. 27:56: Jn. 19:25 "Mary of Clopas"). Jesus himself refers to all of us as being "all brothers" (Mt. 23:8). Other examples of "brothers" used in a non-familial sense are Rom. 14:10, 21; 1 Cor. 5:11; 2 Cor. 8:18; 1 Thess. 4:6; 1 Jn. 3:17; 4:20.

In Gen. 14:14 Lot is described as Abraham's brother, yet Lot is the son of Aran. Lot was Abraham's nephew. Jacob is called the brother of Laban, yet Laban is his uncle (Gen. 29:15). When we look to Dt. 23:7-8 and Jeremiah 34:9 we notice the appellation brothers is used in terms of a person who shares the same culture or national background. In 2 Samuel 1:26 and 1 Kings 9:13 we notice that brother is used in terms of a friend. In Amos 1:9 brother is used as an ally.

"firstborn" (Lk. 2:7)
Ex: 11:5: "every first-born child shall die"—no implication of further births guaranteed
Gen. 27; Ex. 13:2; Nm. 3:12-13; Dt. 21:15-17: firstborn referred to rights and privileges
Ex. 13:2; Nb. 3:12: firstborn referred to the opening of the womb of a woman
Ex.13: 1-16; 34:20: firstborn to be consecrated to God; firstborn referred to being sanctified
Zech. 12:10: "only son" and "first born"
Gen. 49:3: firstborn as preeminent in pride and power
Gen. 12:15-17: birthright as a double share of the father's property which cannot be denied
Ps. 89:28: divine protection and promise

Ex. 4:22: Nation of Israel as "firstborn"
Rom: 8:29: Jesus, firstborn among many (future Christians)
Col. 1:15: Firstborn of all creation
Col. 1:18: Firstborn of new creation
Rev. 1:5: firstborn of the dead, and the ruler of the kings on earth.

"until" (Mt. 1:25)

heos-hou: this compound word makes no further implications

Lk. 1:80: John in desert until day of his manifestation (John remained in the desert after Jesus' manifestation)

Mt. 28:20: I am with you until the end of the age (Jesus will not cease to be with us after the end of the age)

1 Tim. 4:13: until I arrive, attend to reading, teaching... (they obviously will not stop reading or teaching after Paul's arrival)

1 Cor. 15:25: he must reign until he has put his enemies under his foot (Will he cease to reign after this? Obviously not!)

Jn. 19:26-27: unheard of in Jewish culture to give your mother into the care of a non-family member

Mary, as Mother of God and Perpetual Virgin (Two Fathers of Protestantism affirm!!!)

John Calvin, the second most famous Protestant founder, recognized this reality when he stated:

It cannot be denied that God in choosing and destining Mary to be the Mother of his Son, granted her the highest honor.... Elizabeth called Mary the Mother of the Lord because the unity of the person in the two natures of Christ was such that she could have said that the man engendered in the womb of Mary was at the same time the eternal God (*Calvini Opera, Corpus Reformatorum*, Braunschweig-Berlin, 1863-1900, v. 45, 348, 35).

Even Martin Luther, the founder of Protestantism, recognized the important role of Mary as the Mother of God. As he stated in defense of his strong devotion to Mary:

*Mary was made the **Mother of God**, giving her so many great things that no one could ever grasp them...* (*The Works of Luther*, Pelikan, Concordia, St. Louis, v. 7, 572).

What ever happened to his Protestant disciples?

Assumption

Psalm 16:10: the beloved will not know decay

Lk. 1:28: since Mary was full of grace, she bore no original sin and therefore could not experience decay

Gn. 5:24; Heb. 11:5: If Enoch was taken up to heaven, why not Mary?

2 Kg. 2:11: If Elijah was taken up to heaven, why not Mary?

Mt. 27:52: If many saints who had fallen asleep were resurrected "caught up" (in some translations) to meet the Lord in the air, why not Mary?

Rev. 12:1f: Mary clothed with the sun

Tradition about the Assumption

From Gregory of Tours (ca. 590) (1,4, *In Gloria martyrum)*

The course of this life having been completed the Blessed Mary, when called from this world [after having lived a life without original or personal sin], was met by the apostles at her house. When they had heard that she was about to be taken from the world, they kept watch together with her. And behold, the Lord Jesus came with His angels, and taking her soul, He gave it over the Angel Michael and withdrew. At daybreak, the Apostles took up her body on a bier and placed it in a tomb; and they guarded it, expecting the Lord to come. And behold, again the Lord stood by them; and the holy body having been received, He commanded that it be taken in a cloud into paradise: where now, rejoined to the soul, Mary rejoices with the Lord's chosen ones, and is in the enjoyment of the good of an eternity that will never end.

Saints

Communion of saints

Rom. 8:35-39: death cannot separate us from Christ. Therefore, death cannot separate us from his body, the Church. Christ is the head, his body is the Church. The head and body are inseparable in this world and in heaven (cf. Eph. 1:22-23; 5:21-32; Col. 1:18, 24; Rom. 12:5)

Col. 1:24: Make up what is lacking in the sufferings of Christ...

Mk. 9:4: example of communion of saints

Intercessory power of saints

Tob. 12:12: angel presents Tobit and Sarah's prayer to God

Rev. 5:8: elders offer prayers of the holy ones to God

Mk. 9:4: Jesus conversing with Moses and Elijah

Rev. 6:9-11: martyrs under altar seek earthly vindication

Sir. 48:14: in death he did marvelous deeds

Veneration of saints
1 Thess. 1:5-8: you became an example to all believers
Heb. 13:7: imitate the faith and life of leaders
Jos. 5:14: Joshua fell prostrate in veneration before angel
Dan. 8:17: Daniel fell prostrate in terror before Gabriel
Tob. 12:16: Tobiah and Tobit fall to ground before Raphael
Mk. 9:4: build three tents

Relics of saints
2 Kgs. 13:20-21: contact with Elisha's bones restored life
Acts 5:15-16: cures performed through Peter's shadow
Acts 19:11-12: cures through face cloths that touched Paul

Statues (Ex. 20:4-5 prohibition)
Ex. 25:18-19: make two cherubim of beaten gold
Nb. 21:8-9: bronze serpent on pole
1 Kgs. 6:23-29: temple had engraved cherubim, trees, flowers
1 Kgs. 7:25-45: temple had bronze oxen, lions, pomegranates

Tradition

Let us not forget those who have died in our prayer. Let us not forget the patriarchs, prophets, apostles, and martyrs who bring our petitions to God; let us not forget the holy fathers and bishops who have died as well as all those most close to us who bring our petitions to God.

Cyril of Jerusalem (ca. 350)
Catechetical Lectures, 23 [Mystagogic 5], 90

The Last Things

Purgatory
Wis. 3:1-8: gold in a furnace
Rev. 21:27: nothing unclean enters into heaven
1 Cor. 3:15: he will be saved but only as through fire
Mt. 12:36: account for every word on judgment day
Mt. 5:26: will not be released until the last penny is paid
1 Cor. 15:29-30: people being baptized for the dead
2 Tim. 1:16-18: Paul prays for dead friend Onesiphorus
1 Pet. 3:18-20; 4:6: Jesus preached to spirits in prison

Mt. 12:32: sin against Holy Spirit unforgiven in this age or in the age to come

2 Macc. 12:44-46: atoned for dead to free them from sin

Temporal punishment
2 Sam. 12:13-14: David, though forgiven, still punished for sin
Nb. 20:12: Moses would not enter the promised land

Indulgences:
Rom. 12:4-8: Body of Christ
Col. 1:24: make up what is lacking in sufferings

Tradition

Augustine of Hippo in 387 records in his masterpiece *Confessions* the words of his mother, Monica: "All I ask of you is that wherever you may be you will remember me at the altar of the Lord." In other words, prayers were to be said for her, the prayer of the Mass.

In Augustine's *De fide, spe, caritate liber unus* (39, 109) we read:

The time which interposes between the death of a man and the final resurrection holds souls in hidden retreats, accordingly as each is deserving of rest or of hardship, in view of what it merited when it was living in the flesh. Nor can it be denied that the souls of the dead find relief [in purgatory] through the piety of their friends and relatives who are still alive, when the Sacrifice of the Mass is offered for them, or when alms are given in the church. But these things are of profit to those who, when they were alive, merited that they might afterwards be able to be helped by these things. For there is a certain manner of living, neither so good that there is no need of these helps after death, nor yet so wicked that these helps are of no avail after death. There is, indeed, a manner of living so good that these helps are not needed [in heaven], and again a manner so evil that these helps are of no avail [in hell], once a man has passed from this life.

It is not until the 16th century Protestant Reformation that prayers for the dead become seriously challenged.

Being saved
I have been saved (if in a state of grace I am currently saved at this moment in time)

Eph. 2:5-8: by grace you have been saved through faith
2 Tim. 1:9: he saved us, called us, according to his grace
Tit. 3:5: he saved us through a bath of rebirth

No assurance of salvation

Mt. 7:21: not everyone who says "Lord, Lord," will inherit
Mt. 24:13: those who persevere to the end will be saved
Phil. 2:12: work out your salvation in fear and trembling
1 Cor. 9:27: drive body for fear of being disqualified
1 Cor. 10:11-12: those thinking they are secure, may fall
Gal. 5:4: separated from grace, you have fallen from grace
2 Tim. 2:11-13: must hold out to the end to reign with Christ
Heb. 6:4-6: describes those who have fallen
Heb. 10:26-27: if sin after receiving truth, judgment remains
1 Pet. 1:9: as you attain the goal of your faith, salvation
Mt. 10:22: he who endures to the end will be saved
Rom. 13:11: salvation is nearer now than when you first believed
1 Cor. 3:15: he will be saved but only as through fire

Faith and works (inseparable)

Jam. 2:24: a man is justified by works and not by faith alone
Jam. 2:26: faith without works is dead
Gal. 5:6: only thing that counts is faith working in love
1 Cor. 13:2: faith without love is nothing
Jn. 14:15: if you love me, keep my commandments
Mt. 19:16-17: if you wish to enter into life, keep my commandments
Phil. 2:12: work out your salvation with fear and trembling.
Mt. 7:21: not everyone who says "Lord, Lord," will inherit
Rom. 2:2-8: eternal life by perseverance in good works
Eph. 2:8-10: we are created in Christ for good works
Rom. 2:5-8: God will repay each one according to his works
2 Cor. 11:15: their end corresponds to deeds
1 Pet. 1:17: God judges according to one's works
Rev. 20:12-13: dead judged according to their deeds
Col 3:24-25: will receive due payment for whatever you do

Tradition

The *Didache, The Teaching of the Twelve Apostles,* 16, reminds us that we need to "endure in our faith in order to be saved." And the *Epistle of Barnabas* (ca. 96) reminds us that salvation can be lost at any moment:

Let no assumption that we are among the called ever tempt us to relax our efforts or fall asleep in our sins; otherwise the Prince of Evil will obtain control over us, and oust us from the kingdom of the Lord (Barnabas, 4).

If the human race does not have the power of a freely deliberated choice in fleeing evil and in choosing good, then men are not accountable for their actions, whatever they may be. That they do, however, by a free choice, either walk upright or stumble, we shall now prove. God did not make man like the other beings, the trees and the four-legged beasts, for example, which cannot do anything by free choice. Neither would man deserve reward or praise if he did not of himself choose the good; nor, if he acted wickedly, would he deserve punishment, since he would not be evil by choice, and could not be other than that which he was born. The Holy Prophetic Spirit taught us this when he informed us through Moses that God spoke as follows to the first created man: 'Behold, before your face, the good and the evil. Choose the good.'

Justin Martyr (ca. 100)
First Apology, 43

Hell
Mt. 13:49-50; 25:33-46: those who go to hell
Mt. 25:41; 2 Thess. 1:6-9: eternal punishment

Heaven
Mt. 5:3-12; 22:32; 25:33-40; Rom. 8:17: reality

Particular Judgment
Lk. 23:43; 2 Cor. 5:8; Phil. 1:23-24; Heb. 9:27: soul judged immediately upon death

Parousia: Time of second coming unknown
Mt. 24:44: be prepared, Jesus coming at unexpected hour
Mt. 25:13: stay awake, you know neither the day nor hour
Lk. 12:46: master will come at unexpected day and hour
1 Thess. 5:2-3: day of the Lord will come like a thief in the night
2 Pet. 3:9-10: day of the Lord will come like a thief
Rev. 3:3: if not watchful, will come like a thief
Mt. 24:36: no one but Father alone knows day and hour

Last Judgment
Acts 24:15; Jn. 5:28-29: body reunited in a glorified form with our souls—rapture is for those still living when Christ returns. They will

be judged and go eternally with God, **raptured** into heaven, or sent to hell.

Phil. 3:20-21: glorified resurrected body

New Heaven and a New Earth
2 Pet. 3:13; Rev. 21:1-4

Tradition

Premillennialism, Postmillennialism, Amillennialism and the Rapture

Revelation 20 and 1 Thessalonians 4:15-17 are among the most fascinating passages in Scripture in that they have been interpreted in such radically different ways. The primary reason for this is the confusion over the placing of the thousand year reign, the rapture, and over the sense in which these passages were meant to be understood.

Premillennialism

Premillennialism holds that after the period of the Church there will be a time of tribulation that will be followed by Christ's Second Coming, the binding of Satan, and the resurrection of the faithful who have died in Christ. Christ and the risen faithful will reign on earth physically for a thousand years. This will be followed by another period of tribulation, albeit short, the Final Judgment, and the rapture of the faithful into heaven. The creation of a new heaven and a new earth will follow.

Postmillennialism

Postmillennialism holds to the idea that after the period of the Church, Satan will be bound, and a thousand year reign will follow, followed by the rapture into heaven of the living faithful before the period (or during the middle) of the tribulation. This will be followed by the Second Coming of Christ, the resurrection of the dead, the Final Judgment, and the creation of a new heaven and a new earth.

Amillennialism

Catholicism rejects both Premillennialism and Postmillennialism. It believes in what is called Amillennialism. It holds that Revelation 20 is a symbolic passage and that the thousand year reign is a symbolic term for the period from Christ's salvific act to the time of Christ's Second Coming. Christ's Second Coming will be preceded by a short tribulation period. Jesus' return will be followed by the resurrection of the dead (Acts 24:15), the Final Judgment (Mt. 25:31, 32, 46; Jn. 5:28-29; 12:49) and the creation of a new heaven and a new earth (Rom. 8:19-23; Eph. 1:10; 2 Pet. 3:13; Rev. 21:1-2, 4-5, 9, 27) How the transformation of a new heaven and a new earth will take place and how it will look like or when the Second Coming will occur is part of the mystery of our faith.

The Catholic Understanding of the Rapture

And what is the Catholic understanding of the rapture as found in 1 Thessalonians 4:15-17? At the second coming of Christ, the dead will be resurrected, the Final Judgment will take place, and those faithful who are still alive when Christ returns—and after the Final Judgment—will go up with the resurrected faithful to meet and be with Christ forever.

The resurrection of the dead at the end of time is a reference to the resurrection of the bodies of the righteous and unrighteous (cf. Acts 24:15). The bodies of the righteous will be reunited in a glorified form (Phil. 3:21) to their souls in heaven; the bodies of the unrighteous will be reunited to their souls in hell. Those still living in body and soul at the end of time will be judged at that time (the Final Judgment) and then follow Christ in body (in a glorified form) and soul into heaven or go body and soul into hell.

APPENDIX II
Ecce Fides Parish Associations

Ecce Fides is a lay association dedicated to a life of Christian virtue, awareness, and witness in the secular world. Its primary emphasis is on the proclamation and protection of the Catholic faith amidst the influences of the negative aspects of our culture. Members are not members of a religious order, do not take vows, and do not live in community. As Catholics in good standing and as spiritual shepherds they make a commitment from the heart to follow the precepts of the Church, to pray, study, and reflect on a daily basis. They make a commitment from the heart to embark on a continual journey of intellectual and spiritual growth within the traditional spirit of the Catholic Church—as expressed in the *Catechism of the Catholic Faith.*

Requirements of an Ecce Fides Association

1. A family unit or a parish gathering, or any group of Catholics dedicated to learning the *Catechism of the Catholic Church.*
2. Groups are to be led by a priest who lives and preaches the spirit of the Catechism.
3. Meetings are to be weekly.

Structure of Weekly Meetings

1. Opening Prayer

    a. Prayer to the Holy Spirit:

Come, Holy Spirit, almighty Sanctifier, God of love, who filled the Virgin Mary with grace, who wonderfully changed the hearts of the apostles, who endowed all Your martyrs with miraculous courage, come and sanctify us. Enlighten our minds, strengthen our wills, purify our conscience, remedy our judgment, set our hearts on fire, and preserve us from the misfortunes of resisting Your inspirations. Amen.

2. How to ask questions:

    a. What struck you from what you read?
    b. What do you think is the meaning of the passage you read?
    c. What relevance does this passage have for us in today's culture?

3. Read out loud a paragraph of the *Catechism of the Catholic Church* (every paragraph in the Catechism has a bold-faced number, i.e., **1, 2...2856**). After the paragraph reading is complete, share your insights and questions regarding what has been read. Use the questions above (2) to guide you. Then proceed to a reading of the next numbered paragraph followed by a discussion. Repeat this process until the one hour mark has been reached, which concludes this part of the meeting. All are encouraged to read ahead before each meeting and prepare questions, insights, etc.

4. **The** Catechism is divided into four parts. Study one part of the Catechism each week, moving from one section to another weekly. For example,

Week 1: Part One: Profession of Faith

Week 2: Part Two: The Celebration of the Christian Mystery

Week 3: Part Three: Life in Christ

Week 4: Part Four: Christian Prayer

Week 5: Part One: Profession of Faith continued

Etc...

5. Closing Prayer: Finish all meetings in prayer.

        a. Our Father

        b. Glory be...

        c. Hail Mary

Symbolism of Ecce Fides Crest

Shepherd: Sign that we follow the Shepherd of Shepherds, Jesus Christ, and that we are called to be shepherds in praying for priests, the Church, and the world.

Sword and Shield: Symbol that reminds us of our individual and communal obligation to promote and protect the faith.

Jerusalem Cross: Symbol of our role in building up the Kingdom of God, the earthly Jerusalem, and our eternal home and destination, the Heavenly Jerusalem.

Fleur de lys: Symbol of the Trinity and of Mary the Mother of God, perfect model of her Son, perfect model of the Church, perfect model of discipleship.

Papal Keys: Symbol of obedience to the successors of St. Peter, the Rock (Kepa) upon whom the gates of hell will never prevail.

Lion: Symbol of Jesus our Lord, Savior, King, and God.

White Crown: Symbol of a dry martyrdom that all Christians are called to live out in life's daily spiritual battles; that is, faithfulness to one's faith till one's death.

Red Crown: Symbol of "blood martyrdom." That is, the willingness to accept all things from God and for God, even to the shedding of one's blood.

Other great books published by Fr. John J. Pasquini under Shepherds of Christ Publications

Authenticity by Fr. John J. Pasquini

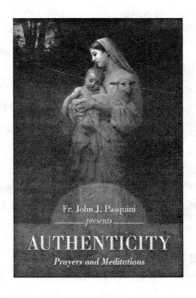

The book of prayers *Authenticity* by Fr. John J. Pasquini is to help one grow ever deeper in the Unitive life.

Apostolic Nuncio –
Archbishop – Philippines

With *Authenticity,* much is gained in prayer, and much is accomplished through prayer. More especially if prayer is directed in behalf of the Church.

Monsignor – Malaysia

I found the book, *Authenticity*, of Fr. Pasquini really good and thought to get copies for some of our priests. May I place an order of 20 copies of this book.

Bishop – Argentina

I received for the second time your letter. The first time it was accompanied with the yellow book *Authenticity*, which I read during the last month of September (with prayers and meditations). I found it very convenient for priests. It did me a lot of good, thanks God.

Bishop – Arlington - USA

I thank you for the copy of *Authenticity*, which was written by Fr. John J. Pasquini. I look forward to using this book of Prayer and I appreciate your thoughtfulness in sending it to me.

Bishop – Ethiopia

The book entitled, *Authenticity, Prayers and Meditations* by Fr. John J. Pasquini is an invaluable source of inspiration for all circumstances of life.

Monsignor – Apostolic Nuncio – Taiwan

I wish to thank you for forwarding this good book, Authenticity, which will help the priests and others to have an intimate communion with the Blessed Trinity.

Bishop – South Africa

I received your book Authenticity with prayers and meditations for all circumstances in our spiritual lives. Thank you sincerely for trying to make people devout. I also thank your society for distributing these books for use by those who try and live spiritual lives.

In Imitation of Two Hearts by Fr. John J. Pasquini

In Imitation of Two Hearts - Prayers for Consolation, Renewal and Peace in Times of Suffering Fr. John J. Pasquini leads a suffering soul to the gentle Hearts of Jesus and Mary. In these most loving Hearts - the prayers by Fr. John Pasquini - help the person suffering to know more deeply the pascal mystery of death/resurrection.

President of the Pontifical Council for Health Pastoral Care - Cardinal - Vatican City

Your publication, *In Imitation of Two Hearts,* will be well used by Pastors to bring consolation to the suffering Brothers and Sisters, I avail myself of this opportunity to renew to you my sentiments of high esteem.

Apostolic Nuncio – Belgium

Many thanks for your kind letter with the useful information and for the consoling book, *In Imitation of Two Hearts.*

Monsignor – Uganda

Thank you for the powerful book of prayer published by Fr. John J. Pasquini *In Imitation of Two Hearts*.

Archbishop – USA

I wish to express my gratitude to you for your kindness in sending me a copy of a special gift of prayers from the book: *In Imitation of Two Hearts* by Fr. John J. Pasquini. I will look at these very carefully and I thank you for them.

Bishop – USA

I thank you for the copy of *In Imitation of Two Hearts*, which was written by Fr. John J. Pasquini. I look forward to using this book of prayers and I appreciate your thoughtfulness in sending it to me.

Cardinal – Japan

Thank you for your kind gift of Christmas season *In Imitation of Two Hearts*. Hope this small book will reach to many people in this time of suffering.

Light, Happiness and Peace by Fr. John J. Pasquini

The purple book *Light, Happiness and Peace* by Fr. John J. Pasquini is a gift and a very strong companion to the prayers *Authenticity* a book of prayers by Fr. John.

This book *Light, Happiness and Peace* is a journey into the spiritual life — an awakening of deeper life IN HIM.

Here are some of the comments we received from bishops and cardinals about the book.

Cardinal – Pontifical Council for Culture – Vatican City

I am sure that this book, *Light, Happiness and Peace* through a discussion on traditional Catholic Spirituality will contribute in bringing back prayer into the mainstream of life.

Bishop – India

Thank you for the book *Light, Happiness and Peace.* The thoughts presented by Fr. John J. Pasquini are very inspiring and will help me in my prayers and talks and sermons to the people.

Prefetto – Vatican City

Received the copy of *Light, Happiness and Peace,* Journeying Through Traditional Catholic Spirituality, by Fr. John J. Pasquini, that you kindly sent to me for the Vatican library. I gladly take this opportunity to express my gratitude for the books and for your kind attention to our Institution.

Patriarch of Egypt

Many thanks for the rich and precious book *Light, Happiness and Peace.*

Archbishop – Uganda

With many thanks, I acknowledge receipt of the special book *Light, Happiness and Peace,* on the spiritual life by Fr. John Pasquini. Your wonderful work to spread devotion to the Hearts of Jesus and Mary is commendable.

Archbishop – Sri Lanka

I am very grateful to you for sending me the special book on the Spiritual life, *Light, Happiness and Peace* by Fr. John Pasquini – which was a sequel to the *Authenticity* book you sent me last August. I am pleased to note that the focus in both those works is on the power of the Eucharist and on devotion to the Hearts of Jesus and Mary.

Medicine of Immortality by Fr. John J. Pasquini

Here are some of the comments we received from cardinals, bishops and a priest about the book *Medicine of Immortality*:

"Fr. John Pasquini's *Medicine of Immortality* is a wonderful source of inspiration for priests, and all who read it, to gain a deeper appreciation of the healing power of the Eucharist. His clear, succinct presentation of the Mass offers a pastorally insightful explanation of the mystery we believe, we celebrate and which we are called to live out in our lives. The prayers and meditations compiled in his book offer opportunities for spiritual reflection which will assist the reader in growth toward a deeper understanding of the mystery of the Eucharist."

Anthony Cardinal Bevilacqua,
Archbishop Emeritus of Philadelphia

"In *Medicine of Immortality*, Father John Pasquini offers his readers the richness of Catholic devotional prayer, the wisdom of the Fathers and, most of all, the fruits of his own prayer and meditation before the Blessed Sacrament. I recommend this book to all who wish to grow in their love for the Lord, who sustains the life of His Church through the precious gift of His Body and Blood."

Francis Cardinal George,
O.M.I., Archbishop of Chicago

"The work of Father John Pasquini in writing the book *Medicine of Immortality* is evidence of his own great love for Christ in the Holy Eucharist, and of his determination to teach what Christ has taught and the Church has reiterated since her beginning."

The Most Reverend Fabian W. Bruskewitz,
Bishop of Lincoln, Nebraska,
author of *The Catholic Church:*
Jesus Christ Present in the World.

"The Apostolic Exhortation of our Holy Father, Pope Benedict XVI, *Sacramentum Caritatis*, is a great gift to the Church presenting a true compendium on the sublime Sacrament of the Eucharist. In accord with the spirit of Eucharistic reverence which the Pope's words inspire, Father Pasquini's work, *Medicine of Immortality*, is a reflection of the writer's personal devotion to this Sacrament of Sacraments. By sharing his compilation of prayers and meditations, Father Pasquini speaks to the mystery that the awesome depths of the Eucharist are unfathomable."

Most Reverent Gerald M. Barbarito,
Bishop of Palm Beach

"I pray that Fr. Pasquini's book will inspire great devotion to the Eucharist among the faithful on a very wide scale. He has provided us with all the resources we need both for correct understanding of this greatest of Gifts and for fervent devotion. I will remain forever grateful to him for this truly wonderful priestly and apostolic work. In pointing people directly to the source of all healing and immortality, he is just doing what priests do best: witness Christ!"

Rev. Fr. Thomas J. Euteneuer,
columnist,
President, Human Life International,
EWTN Contributor

Shepherds of Christ Associates

The Shepherds of Christ has prayer chapters all over the world praying for the priests, the Church and the world. These prayers that Father Carter compiled in the summer of 1994 began this worldwide network of prayer. Currently the prayers are in six languages with the Chuch's *Imprimatur*. Fr. Carter had the approval of his Jesuit provincial for this movement, writing the Newsletter every 2 months for 6 1/2 years. After his death, and with his direction, we in the Shepherds of Christ circulated the *Priestly Newsletter Book II* to 95,000 priests with other writings. We have prayed daily for the priests, the Church, and the world since 1994.

Associates are called to join prayer Chapters and help us circulate this newsletter centered on spreading devotion to the Sacred Heart and Immaculate Heart and helping to renew the Church through greater holiness. Fr. John J. Pasquini is a tremendous gift on this 12th anniversary year of the newsletter. Form a Prayer Chapter & order a Prayer Manual.

Also we have people who spend 2 hours weekly before the tabernacle praying for these intentions:

1.) For the spread of the devotion to the Hearts of Jesus and Mary culminating in the reign of the Sacred Heart and the triumph of the Immaculate Heart.
2.) For the Pope.
3.) For all bishops of the world.
4.) For all priests.
5.) For all sisters and brothers in the religious life.
6.) For all members of the Shepherds of Christ Movement, and for the spread of this movement to the world.
7.) For all members of the Catholic Church.
8.) For all members of the human family.
9.) For all souls in purgatory.

This movement, *Apostles of the Eucharistic Heart of Jesus,* was begun by Fr. Carter. Please inquire. Shepherds of Christ Ministries P.O. Box 627, China, Indiana 47250 USA 1-888-211-3041 or (812) 273-8405 Fax: (812) 273-3182 E: info@sofc.org